Philosophy of Biology Before Biology

The use of the term "biology" to refer to a unified science of life emerged around 1800 (most prominently by scientists such as Lamarck and Treviranus, although scholarship has indicated its usage at least 30–40 years earlier). The interplay between philosophy and natural science has also accompanied the constitution of biology as a science.

Philosophy of Biology Before Biology examines biological and proto-biological writings from the mid-eighteenth century to the early nineteenth century (from Buffon to Cuvier; Kant to Oken; and Kielmeyer) with two major sets of questions in mind:

1 What were the distinctive conceptual features of the move toward biology as a science?
2 What were the relations and differences between the "philosophical" focus on the nature of living entities, and the "scientific" focus?

This insightful volume produces a fresh but also systematic perspective both on the history of biology as a science and on the early versions of, in the 1960s in a post-positivist context, the philosophy of biology. It will appeal to students and researchers interested in fields such as history of science, philosophy of science and biology.

Cécilia Bognon-Küss is a postdoctoral researcher at the University Paris Diderot, France.

Charles T. Wolfe is a researcher in the Department of Philosophy and Moral Sciences, Ghent University, Belgium.

History and Philosophy of Biology

Series Editor: Rasmus Grønfeldt Winther is Associate Professor of Philosophy at the University of California, Santa Cruz (UCSC).

This series explores significant developments in the life sciences from historical and philosophical perspectives. Historical episodes include Aristotelian biology, Greek and Islamic biology and medicine, Renaissance biology, natural history, Darwinian evolution, nineteenth-century physiology and cell theory, twentieth-century genetics, ecology, and systematics, and the biological theories and practices of non-Western perspectives. Philosophical topics include individuality, reductionism and holism, fitness, levels of selection, mechanism and teleology, and the nature/nurture debates, as well as explanation, confirmation, inference, experiment, scientific practice, and models and theories vis-à-vis the biological sciences.

Authors are also invited to inquire into the "and" of this series. How has, does, and will the history of biology impact philosophical understandings of life? How can philosophy help us analyze the historical contingency of, and structural constraints on, scientific knowledge about biological processes and systems? In probing the interweaving of history and philosophy of biology, scholarly investigation could usefully turn to values, power, and potential future uses and abuses of biological knowledge.

The scientific scope of the series includes evolutionary theory, environmental sciences, genomics, molecular biology, systems biology, biotechnology, biomedicine, race and ethnicity, and sex and gender. These areas of the biological sciences are not silos, and tracking their impact on other sciences such as psychology, economics and sociology, and the behavioral and human sciences more generally, is also within the purview of this series.

Recent Titles

Revisiting the *Origin of Species*
The Other Darwins
Thierry Hoquet

Philosophy of Biology Before Biology
Edited by Cécilia Bognon-Küss and Charles T. Wolfe

For more information about this series, please visit: www.routledge.com/History-and-Philosophy-of-Biology/book-series/HAPB

Philosophy of Biology Before Biology

Edited by Cécilia Bognon-Küss and Charles T. Wolfe

Routledge
Taylor & Francis Group
LONDON AND NEW YORK

First published 2019
by Routledge
2 Park Square, Milton Park, Abingdon, Oxon OX14 4RN

and by Routledge
52 Vanderbilt Avenue, New York, NY 10017

Routledge is an imprint of the Taylor & Francis Group, an informa business

British Library Cataloguing-in-Publication Data
A catalogue record for this book is available from the British Library

Library of Congress Cataloging-in-Publication Data
A catalog record has been requested for this book

ISBN: 978-1-138-65287-3 (hbk)
ISBN: 978-1-315-62395-5 (ebk)

Typeset in Times New Roman
by Wearset Ltd, Boldon, Tyne and Wear

This volume is affectionately and respectfully dedicated to the memory of Jean Gayon (1949–2018)

Contents

Contributors

Cécilia Bognon-Küss is a postdoctoral researcher at the University of Paris 7 ("Who am I?" Labex) and at the IHPST (CNRS), working on "The classical concept of metabolism, biological identity and the challenges from microbiome research." Her dissertation (at the Institut d'Histoire et de Philosophie des Sciences et des Techniques, Université Paris 1 Panthéon-Sorbonne) was entitled "Between Biology and Chemistry: Nutrition, Organization, Identity" and examined the role played by the investigation of nutrition, and the constitution of the concept of metabolism, in the emergence of biology in the eighteenth and nineteenth centuries. She edited a special issue of *History and Philosophy of the Life Sciences* on the topic "Organic – Organization – Organism: Essays in the History and Philosophy of Biology and Chemistry" (2018).

Tobias Cheung is Associate Professor for the History of Knowledge and Culture at the Institute of Cultural History and Theory of the Humboldt University in Berlin. He is the author of *The Organization of the Living: The Role of Cuvier's, Leibniz's and Kant's Notions of Organismic Order in the History of Biology* (2000), *Charles Bonnet's System Theory and Philosophy of Organic Order* (2005), *Res vivens. Agent Models of Organic Order 1600–1800* (2008) and *Organisms. Agents between Inner and Outer Worlds 1780–1860* (2014), and editor of *Transitions and Borders between Animals, Humans and Machines 1600–1800* (2010).

François Duchesneau is a Professor Emeritus at the Université de Montréal. His publications include: *Leibniz et la méthode de la science* (1993); *La Dynamique de Leibniz* (1994); *Philosophie de la biologie* (1997); *Les Modèles du vivant de Descartes à Leibniz* (1998); *Leibniz. Le vivant et l'organisme* (2010); *La Physiologie des Lumières* (2012) and (with J. E. H. Smith), *The Leibniz-Stahl Controversy* (2016). His present research bears on the concept of organism in modern life sciences.

Snait Gissis teaches at the Cohn Institute for the History and Philosophy of Science and Ideas, Tel Aviv University. In recent years she has been working on the interactions between the social and the biological, in particular Lamarckisms and the constitution of the emerging social sciences at the end of the nineteenth century (forthcoming in *Studies in History and Philosophy of the Biological and Biomedical Sciences*: "Transfer of Lamarckisms and Emerging 'Scientific' Psychologies: 19th–Early 20th Centuries Britain and France"), the emergence of "race" as a scientific object and category in the late eighteenth century and its continued deployment until the very present at the interface of the life sciences, medicine and society, and the notion of collectivity (recently coediting, with Ehud Lamm and Ayelet Shavit, *Landscapes of Collectivity in the Life Sciences*, 2018).

Ina Goy is Senior Lecturer in Philosophy at the University of Tübingen (on leave) and Professor in the history of Western philosophy at Beijing Normal University. She is author of the monograph *Kants Theorie der Biologie* (2017) and coeditor of the anthology *Kant's Theory of Biology* (2014). She is currently working on an edited volume, *Kant on Proofs of the Existence of God* and on a collaboratively written commentary *Aristotle. The Generation of Animals*. Her recent articles in the history and philosophy of biology include "Was Aristotle the 'Father' of the Epigenesis Doctrine?" (2018), "Kant on Human and Non-Human Animals" (2018), "The Antinomy of Teleological Judgment" (2015) and "Kant's Theory of Biology and the Argument from Design" (2014).

Philippe Huneman is a research director at the Institut d'Histoire et de Philosophie des Sciences et des Techniques (CNRS/Université Paris I Sorbonne). A philosopher of biology, he works on issues related to evolutionary biology, on the concept of organism and its relations to Kant's metaphysics, and on kinds of explanation in biology and ecology (with a focus on structural/topological explanations). The author of *Métaphysique et biologie* (on Kant and the concept of organism, 2008), to be translated and published soon by Routledge, he has edited several books, including *Understanding Purpose: Essays on Kant's Philosophy of Biology* (2007), *Functions: Selection and Mechanisms* (2013), *From Groups to Individuals* (2013, with F. Bouchard), *Challenging the Modern Synthesis* (2017, with D. Walsh) and *Handbook of Evolutionary thinking in the Sciences* (2015, with T. Heams, G. Lecointre and M. Silberstein). He is coeditor of the book series History, Philosophy, and Theory in Life Sciences.

Lynn K. Nyhart is Vilas-Bablitch-Kelch Distinguished Achievement Professor in the Department of History at the University of Wisconsin-Madison. She is the author of *Biology Takes Form: Animal Morphology and the German Universities, 1800–1900* (1995) and *Modern Nature: The Rise of the Biological Perspective in Germany* (2009), and coedited (with Scott Lidgard) *Biological Individuality: Integrating Scientific, Philosophical, and Historical Perspectives* (2017). Currently she is working on a project on the relationship of German life scientists to the revolutions of 1848 and their aftermath.

Stéphane Schmitt is a research director at the French Centre National de la Recherche Scientifique (Laboratoire SPHERE, Paris). He works on the history of the life sciences, especially in the eighteenth and nineteenth centuries. He has published books and papers on the history of anatomy, embryology and the sciences of evolution, and is the main editor of Buffon's *Œuvres Completes*, published by Honoré Champion (Paris, 2007, 12 volumes published to date).

Phillip Sloan is Professor Emeritus in the Program of Liberal Studies and the Graduate Program in History and Philosophy of Science program at the University of Notre Dame. His research area is the history and philosophy of the life sciences in the modern period, including the history of evolutionary theory, Enlightenment natural history, and intellectual history. He has also worked in recent years in the history and philosophy of recent genetics and molecular biology. Sloan's most recent books include *Creating a Biophysics of Life: The Three-Man Paper and Early Molecular Biology* (Chicago, 2012), and he is main editor and contributor to *Darwin in the Twenty-First Century* (UND Press, 2015). He is currently working on a major book project, *Mastering Life*. He is also involved in the first full English translation of Émilie Du Châtelet's *Institutions de Physique* (1740), which will appear in 2019.

Georg Toepfer is Head of the Knowledge of Life department at the Centre for Literary and Cultural Research (ZfL) in Berlin. His principal area of research is the history and philosophy of the life sciences, with a special focus on the history of biological concepts and their transfer between biology and other fields. Major publication: *Historisches Wörterbuch der Biologie. Geschichte und Theorie der biologischen Grundbegriffe* (3 vols, 2011).

Charles T. Wolfe is a researcher in the Department of Philosophy and Moral Sciences, Ghent University, working primarily in history and philosophy of the early modern life sciences, with a particular interest in materialism and vitalism. He is the author of *Materialism: A Historico-Philosophical Introduction* (2016) and the forthcoming monograph *La philosophie de la biologie: une histoire du vitalisme* (Garnier) and has edited volumes including *Monsters and Philosophy* (2005), *The Body as Object and Instrument of Knowledge* (2010, with O. Gal), *Vitalism and the Scientific Image in Post-Enlightenment Life-Science* (2013, with S. Normandin), *Brain Theory* (2014) and *Physique de l'esprit* (with J. C. Dupont and C. Cherici, 2018). He also coedits the Springer series in History, Philosophy and Theory of the Life Sciences. Papers and other works are available at http://ugent.academia.edu/CharlesWolfe.

John H. Zammito is John Antony Weir Professor of History at Rice University. He works in history of philosophy and science, concentrating on the German Enlightenment, especially Kant and Herder, and the emergence of life sciences in eighteenth-century Germany. Publications include: *The Genesis of Kant's Critique of Judgment* (1992), *Kant, Herder and the Birth of Anthropology* (2002), *A Nice Derangement of Epistemes: Post-Positivism in the Study of Science from Quine to Latour* (2004) and *The Gestation of German Biology: Philosophy and Physiology from Stahl to Schelling* (2018).

Acknowledgments

The "ABC model of flower organ identity" was originally published in 2005 ("From leaf to flower: Revisiting Goethe's concepts on the 'metamorphosis' of plants," *Brazilian Journal of Plant Physiology*, 17.4). We would like to thank the authors, Marcelo and Odair Dornelas, and the editors-in-chief, Ricardo Bressan-Smith and Gustavo Habermann, for the reprint permission of the drawing. The "metamorphosis of foliage leaves of a sow-thistle" was originally published in 1964 ("Der Pflanzentypus als Bewegungsgestalt. Gesichtspunkte zum Studium der Blattmetamorphosen," *Elemente der Naturwissenschaft*, 1), the "transformation of the leaves of a blooming shoot of the wild peony" in 1970 ("Staubblatt und Fruchtblatt. Beiträge zum Verständnis der Bildebewegung im Blütenbereich," *Elemente der Naturwissenschaft*, 13). We would like to thank the author, Jochen Bockemühl, and the Anthroposophical Society at the Goetheanum, especially Ruth Richter and Mara Born, for the reprint permissions of both drawings, which have been slightly modified for modern reprinting.

We would like to thank Rasmus Winther for his interest in our project, and Elena Chiu at Routledge for her persistence in following its initial vagaries.

Introduction

Cécilia Bognon-Küss and Charles T. Wolfe

A volume that speaks of 'philosophy of biology before biology' may seem at first glance (and perhaps even later) to be a piece of unrepentant anachronism, if not a return to the bad Whiggish habits of several generations earlier. What could it mean to speak of philosophy of biology before biology itself – literally, a philosophical discipline that emerged in the 1960s presented as somehow 'before' a science that emerged in the late eighteenth century? In fact, we suggest that the philosophy of biology existed in a variety of forms before the particular professional crystallization we are familiar with today (and which was one of several competing 'paradigms' at the time, including more holistic projects such as 'biological philosophy').

However, if we accept to philosophize about a discipline that did not yet exist, and that this project is not tied to the historical constraint of the existence of the discipline of biology as such, we could symmetrically question the choice of what should count within this 'biology before biology'. That is, the global project of a 'philosophy of biology before biology' might conversely seem artificial regarding its periodization: why force-fit all this into the eighteenth and nineteenth centuries? One could argue for instance that the Scientific Revolution might have had interesting 'philosophy of biology before biology' outcomes. However, if we enlarged our content to include this period, because early modern medicine (which is a somewhat different topic, with its own historiography etc.) suddenly is part of our conceptual reconfiguration, then why not embrace Descartes (mechanism), Leibniz (organism) or even Aristotle (teleology)? We would say that this is not part of our project because this book focuses on the conceptual conditions of the emergence of an autonomous discipline, not its archaeology.

Biology came together as an integrated science of living functions precisely by integrating different disciplines, including physiology, anatomy and medicine. Issues concerning its 'naming' and various definitions are addressed elsewhere (McLaughlin 2002, and our chapter in the present volume). What is striking here is the constitution of a unified framework to develop investigations focusing on 'vital' phenomena. The philosophy of biology, on its part, comes out of the post-positivist tradition of the philosophy of science in the 1960s. However, philosophers obviously did not wait for the 1960s to raise questions about the fact and the nature of life: as we discuss in our chapter, from major figures such Aristotle, Descartes, Leibniz and Kant to less-known eighteenth-century philosophers and thinkers who might have been called 'natural philosophers' in earlier generations, philosophers have been preoccupied with the nature of life, oftentimes in order to nourish their own versions of a metaphysics of substance, or a living matter theory, or a naturalized monadology …

Vital forces, the human–animal boundary, the creativity and spontaneity of nature, and the transformation of living forms over time became major metaphysical issues and 'scientific' issues, inseparably. Just as some biologists (and philosophers) will say that there is no particular need for a concept of life to somehow 'ground' or 'unify' this science – 'Biology has developed as a discipline without having anything terribly precise to say about exactly what its domain of inquiry is', so that 'perhaps "alive" is not a crisply delimited category in nature' – others in equal numbers will declare that there *is* a need for such a concept, whether it is presented as a endogenous contribution from philosophy or an exogenous, internally grown biological concept: 'The question, "*What is life?*" lay behind everything I learned. Life seemed to be characterized by a peculiar reasonableness and purposefulness of instinctive involuntary action.'[1]

Philosophers moving away from the 'Scientific Revolution' focus on mechanism, laws of nature, and gravity towards teleology, embodiment, organization and vitality understood as a metaphysical problem are equally focused on these shifting scientific grounds, which serve as a backdrop for the constitution of biology as a science. It involved conjointly French, German, Italian, English and Scottish authors, among others. It is characterized by a set of issues, concepts and argumentative positions that we term here 'philosophy of biology before biology'. The current volume intends to explore some of its salient aspects.

One of our leading hypotheses is that such a project is relevant to the philosophy of biology currently defined, especially when it comes to issues like the relations between chemistry and biology, the status of development or the ontological nature of biological organization. One of the specific emphases present in this volume, unlike other studies of the genesis of biology, is the interaction also between 'vital(ist)' and chemical considerations. Our volume is historical in nature, yet it is not (another) history of biology, being rather problem-oriented. It examines these materials – *biological* and proto-biological writings from the mid-eighteenth century to the early nineteenth century (e.g. from Buffon to Cuvier, from Kant to Oken and Kielmeyer) and *philosophical* reflections on life – with two major sets of questions in mind: what were the distinctive conceptual features of the move towards biology as a science? What were the relations but also differences between the 'philosophical' focus on the nature of living entities and the 'scientific' focus, including in overlapping cases such as the notion of vital forces, or that of organism, where metaphysical and empirical issues are very tightly interwoven? Some authors such as Lamarck would argue that all vital properties are due to the emergent consequences of organization; other such as Diderot held complex and changing views on this question. On his part, Kant tried to elaborate a concept of 'organized body' that integrates the requirements of biological judgement. Many other examples could be given to show that 'organization' was a central concern for all discussion about the nature and requirements of biological knowledge in the eighteenth century.

Rather than studying traditional doctrines, i.e. labels like 'mechanism' or 'vitalism' (thinking of the line attributed to Paul Valéry, according to which 'One cannot get drunk with the labels on bottles'[2]), this volume rather concentrates on specific themes and problems, which we have organized into three parts: *Form and Development*, *Organism and Organization*, and *Systems*. After an initial chapter on the idea of 'philosophy of biology before biology' itself (Bognon-Küss and Wolfe), in the first part our contributors discuss different cases of the interplay between conceptual and empirical intentions and dimensions concerning the generation of living forms, in Buffon (S. Schmitt), Émilie Du Châtelet (P. Sloan) and Wolff's reception by Herder, Tetens and Kant (J. Zammito). In the second part, they discuss issues concerning both organism and vital chemistry, in Spallanzani and Senebier (F. Duchesneau) and Cuvier, Hufeland and Cabanis (T. Cheung). In the third and final part, the chapters cover various possible systematic projects that are either coeval with biology or require its 'rethinking' (Toepfer on Kant and ecology, Goy on Goethe, and Gissis on Lamarck and biology). Finally, reflecting the complex nature of the genre of philosophy of biology treated in a historical sense (and, conversely, the way in which the chapters written for this volume approach their historical material with a 'conceptual' focus), we have asked two prominent figures from different ends of the discipline to contribute postscripts on the contents of the volume: Philippe Huneman from the standpoint of a (historicized) philosophy of biology, and Lynn Nyhart from that of a historian of biology.

We believe this volume will produce a fresh but also systematic perspective on both the history of biology as a science and on the early versions of what came to be called the philosophy of biology. Additionally, owing to its conceptual and metaphysical focus, this volume should build a bridge between more historical understandings of these materials, and contemporary philosophy of biology.

Notes

1 Respectively, Sober 2003, 318; Reich 1968, 45.
2 Cit. in Esposito 2013, 8.

References

Esposito, M. 2013. *Romantic Biology, 1890–1945*, London: Pickering & Chatto.

McLaughlin, P. 2002. Naming biology. *Journal of the History of Biology*, *35*: 1–4.

Reich, W. 1942/1968. *The Function of the Orgasm: Sex-Economic Problems of Biological Energy*, trans. T. P. Wolfe, London: Panther.

Sober, E. 2003. Philosophy of biology, in N. Bunnin (ed.), *Blackwell Companion to Philosophy*, 2nd revised edition (pp. 317–344), London: Blackwell.

1 The idea of 'philosophy of biology before biology'

A methodological provocation

Cécilia Bognon-Küss and Charles T. Wolfe

Philosophy of biology before biology: an absurd idea?

The title of this volume and of the present chapter may sound provocative. Scholars may feel their anachronism radar stirring to life: in what sense could there be 'philosophy of biology,' the name for a professional subdiscipline in philosophy which emerged – in its current form – in the 1960s, before the emergence of biology itself, sometime in the late years of the eighteenth century?[1] Indeed, beyond the problem of anachronism, this seems like a matter of common sense. In fact, we believe that philosophy of biology existed in a variety of forms before the particular professional crystallization we have mentioned.

Differently put, philosophers obviously did not wait for the 1960s to raise questions about the fact and the nature of life: Aristotle's *telos*, Descartes's 'animal-machine,' and Kant's 'teleological judgment' immediately come to mind for any contemporary philosopher. More specifically, in the eighteenth century, philosophers interested in empirical findings concerning vital phenomena shared the concerns of naturalists facing recently observed organic phenomena that were difficult to explain: the regeneration of polyps, details of the embryogenesis of chicks, Galvanism, Mesmerism, and so on. This could be illustrated by Diderot's writings on Trembley's polyp in the 1750s–1760s, Charles Bonnet's reactivation of Leibniz's preformationism (1762, 1764, 1769), or Kant's engagement with the question of nervous diseases, in the 1760s but also in his later *Anthropology* (1798).[2] Metaphysical issues such as organization, vital forces and the creativity of nature thus came to the fore. This interplay between philosophy and natural science accompanied the constitution of biology as a science, jointly involving French, German, English and Scottish authors, to name the more prominent cases.

Biology and philosophy of biology

Despite the appearance of more careful specialized scholarship over the past few decades, it is still standard to trace the usage of the term 'biology' back to Treviranus and Lamarck around 1800,[3] but, as Jacques Roger observed nicely, 'biology did not just appear suddenly at the end of the eighteenth century like

Athena born from the head of Zeus.'[4] Earlier, the term 'biology' was used in the context of German *Naturphilosophie*, in Theodor Georg August Roose, Karl Friedrich Burdach, and Carl Christian Erhard Schmid, in the 1790s,[5] and recent research has pointed to its usage in a treatise of natural philosophy by a Danzig professor named Michael Christoph Hanov, the four-volume *Philosophia naturalis sive physica dogmatica*, of which volume 3, published in Halle in 1766, deals with 'geology, biology and botany.'[6] Yet, the exact meaning of the term 'biology' remained unstable until the end of the century (Linnaeus had already used it a generation earlier, but to mean something close to 'biography,' a usage which Kai Torsten Kanz has traced back as far as 1660), and, conversely, if one approaches the issue from the standpoint of the existence of a new science of life and how to designate it, the terminology was quite variable, including 'zoonomy,' 'general zoology,' 'biology,' 'physiology,' 'bionomy,' 'biogeography,' and 'general natural history,' a term which was used until the mid-nineteenth century, e.g., by Isidore Geoffroy Saint-Hilaire.[7] The term 'life science' or 'science of life' (*Lebenswissenschaft*; *Wissenschaft des Lebens*) itself appears in the title of an 1800 work by Christoph Meiners (Meiners 1800), but in an antiquated sense, which refers to morals – the 'art of living,' as it were. Only some 15 years later this expression would be used in the sense of a specifically biological (or medical, or physiological) body of knowledge, in the first sentence of the Chevalier de Richerand's physiology textbook.[8]

It is thus widely agreed among historians that biology as a science of the functioning and development of living bodies emerged at the beginning of the nineteenth century, integrating methodological or empirical advances in various disciplines, namely physiology, embryology, comparative anatomy, natural history, and medicine. The fact that the word 'biology' was simultaneously and independently coined by several authors from different national and disciplinary backgrounds (Hanov 1766, Bichat 1800, Lamarck 1809, Treviranus 1802–1822) is commonly seen as a testimony of this epistemic emergence (Duchesneau 1982, Barsanti 1994, 2000, McLaughlin 2002, Wolfe 2011b). Even though scientists had, of course, been dealing with living phenomena prior to this, what is striking here is the constitution of a unified framework for developing research focused on 'vital' phenomena.

'Philosophy of biology' in turn is a fairly recent area of academic expertise. It developed from the postpositivist tradition of the philosophy of science in the late 1960s,[9] and is characterized by specific journals (*Biology and Philosophy*, *Studies C*, *Biological Theory*…), scientific societies (e.g., ISHPSSB), and a core group of issues that mostly revolve around evolutionary biology and molecular biology and are supposed to be simultaneously relevant for metaphysics and theoretical biology, e.g., molecular reductionism, adaptationism, units of selection, genetic information and so on. Biologists such as Ernst Mayr and Stephen Jay Gould were highly influential in this growing subfield of philosophy. Originally stemming from the desire to emancipate the field from neopositivist philosophy of science and reductionism (Hull 1969), philosophy of biology developed as a specific field devoted to the study of biological science as such, viewed as an autonomous science – and not as a mere exemplification of mainstream philosophy of science concepts and problems. So conceived, philosophy of biology is quite exclusive of historical concerns, even though its 'founding parents' (Michael Ruse, Marjorie Grene, and David Hull) actually devoted a significant amount of their work to the history of biology (see Gayon 2009, *passim*, on the question of the gradual self-definition of philosophy of biology as a discipline).

Biological philosophy and philosophy of biology

One earlier distinction was that between 'biological philosophy' and 'philosophy of biology.' Those two expressions are rather old, respectively coined by Auguste Comte in his *Cours de Philosophie Positive* (1830–1842; volume 3, dealing with 'chemical philosophy and biological philosophy,' appeared in 1838), and William Whewell in *Philosophy of the Inductive Sciences* (1840). For Comte, 'biological philosophy' referred to something like theoretical biology, i.e., the systematic enquiry into biology's 'fundamental conceptions,' and by that he meant that 'biological philosophy' was a genuine part of biology – its theoretical part (Gayon 2009).[10] Contrasting with that view, the 'philosophy of biology' Whewell advocated for was in principle external to biology, and consisted in the reflexive and critical examination of its concepts. Thereon, 'biological philosophy' freed itself from Comte's original intentions, and was commonly used so to designate the attempt to ground the elucidation of traditional philosophical questions about life, the place of man in the universe, the nature of mind and free will, on the bases of current biological knowledge. It predated the 'philosophy of biology' and is instantiated in works such as Kurt Goldstein's *Der Aufbau des Organismus* (Goldstein 1934), Woodger's *Biological Principles* (Woodger 1929), or Hans Jonas's *Phenomenon of Life* (Jonas 1966) or arguably earlier in works on metaphysics such as those of Whitehead or Bergson. 'Philosophy of biology' in its current, academic sense is mostly concerned, as noted above, with conceptual issues in biology as an institutionalized science, thus raising issues proper to this science as such, at the risk of drowning out other equally central topics, as we will elaborate on below.

Yet, this helpful distinction may not be enough to capture the nature of philosophical reflection concerning biological phenomena. For such reflection is neither 'biological philosophy,' because this project is not exactly grounded on an extant biology, with its uncontroversial findings and its identified avenues of enquiries, nor does it target a set of issues emerging from biological practice and theorizing, as 'philosophy of biology' does. To some extent, there is a sense that Buffon's speculations on the originality of organic matter, Kant's theory of organized beings and natural purposes, and many other eighteenth-century scientists'/philosophers' accounts of life, organic matter, generation, etc. that predated the formation of biology as a science belong to this interplay between biological philosophy and philosophy of biology. The present volume intends to consider this interplay, for which we propose the label, 'philosophy of biology before biology.'

Such an interplay has two 'dimensions,' so to speak. There is a sociological dimension, constituted by the fact that at some point philosophy and biology were not separate fields, practiced by different actors in distinct institutions – and this impinges on the content of any philosophical discussions of objects and issues pertaining to 'life.' And there is a dimension proper to the 'conceptual foundations of science' (or what some choose to call 'historical epistemology,' a term we discuss below), constituted by the fact that, whereas some current questions in the philosophy of biology today (e.g., reductionism, information, selection, adaptation, and teleology) are a bit like scientific questions, i.e., they are framed in paradigmatic terms, with rival accounts (is genetic information a spurious concept or not? what is the extent of adaptation in evolution? etc.), this paradigmatic framing of philosophical issues did not exist in the period 'before biology.' Neither of these two dimensions yields a picture of 'philosophy of biology before biology' that is reducible to philosophy of biology, or biological philosophy.

'Historical epistemology of the life sciences' and philosophy of biology

We find the way in which philosophy of biology crystallized around a core set of concepts and problems tied to evolutionary biology to be problematic,[11] namely in the way it does not reflect the wealth and diversity of philosophical problems contained in biology. In that sense, one of the goals of the present volume is to create some distance with regard to the 'problem space' that seems to have been imposed on us by a certain history of biology, with its highlighted or heroic entities such as the organism or the gene.[12] Which philosophical consequences might we expect (or hope for) from a philosophy of biology before biology, that is, a philosophy of biology which is both historically inspired and deliberately anachronistic (without this anachronism implying a privilege of biological concepts as they exist in 2018)?

Even though the philosophy of biology has tended to set aside or ignore the tradition we might term, reflecting current usage, 'historical epistemology of the life sciences,' viewing the latter as an historical rather than philosophical enterprise, this tradition never took itself to be separate from philosophy. With the idea of a French 'epistemological' tradition in the background (which emphasizes the study of the conceptual dimensions of science, not the theory of knowledge as in the English meaning of the term), we speak of a historical epistemology of the life sciences, rather than strictly of their history, because we focus on the epistemic constructions of biology in order to jointly emphasize their specific historicized nature (it's not that Treviranus's 'biology,' Claude Bernard's 'physiology,' or Erasmus Darwin's 'zoonomy' are all the same projects) and their conceptual status (i.e., their philosophical pertinence in understanding biology). The historical epistemology tradition focuses on the emergence of forms of knowledge in given historical contexts but considers this approach to be part and parcel of a philosophical project. Jean Gayon has pointed to the influence of Auguste Comte on this tradition, for which the history of science is not just a legitimately philosophical enterprise but also a crucial one for philosophy. Earlier, Georges Canguilhem had already insisted on the history of science as a philosophical project:

> a theory of knowledge without reference to epistemology would be a meditation on the void … an epistemology without any relation to the history of science would be a wholly superfluous clone of the science which it claims to discuss.[13]

Differently put, the historical epistemology tradition emphasizes that the history of science cannot dispense with theoretical and indeed philosophical dimensions, given that science seeks to develop instruments to approximate 'truth,' or a truthful relation to the phenomena. Yet the historical dimension is there to prevent over-generalization or abstraction from particular contexts.[14]

To address biology historically while asking conceptual questions in this case is not just a 'history of ideas' approach (like in the older studies of the 'great chain of being' or 'evolution, the history of an idea'). For our approach explicitly inquires into the origins and emergence of biology, *as a way of challenging a certain univocity and unidirectionality*: it is a way of attending to the genealogy of biology 'before biology,' to the various strata of meanings and the problem spaces that this science contains. The potentially anachronistic dimension of our usage of the term 'biology,' in this chapter but also by different authors in this volume, also takes inspiration from the philosophical dimension in historical epistemology, which Jean Gayon, writing about the work of the French historian of the life sciences Jacques Roger, described as 'forcing the concept to show itself, in the history of science.'[15] Such a genealogy should then yield properly philosophical effects, reflecting our claim that the problems defining and shaping the philosophy of biology are – at least in part – inherited from the way the narrative of its emergence was constructed. The main trend in the studies on the historical and conceptual conditions of the emergence of biology as a discipline has been to highlight the centrality of the notion of organism, organized beings, or organized wholes (Duchesneau 1982, 2018, Cheung 2000, 2006, Huneman 2008, Wolfe 2010). But this comes at the cost of leaving aside, more or less, the question of the difference and relation between organic and inorganic matter, and the emergence of organization per se (questions which are treated in several of the chapters of the present volume). To be sure, the organism concept is crucial in the elaboration of modern biology, whether this is specifically in embryology (the development of living forms through epigenesis), pre-Darwinian evolution (transformist explanations of living forms), or comparative anatomy (Sloan 2003, Huneman 2006). Yet, the distinction between organized (or organic) and inorganic bodies, and the efforts to provide a molecular and chemical account of organic function, are equally crucial motivations of and contributions to biology as a science.[16] These are instances of different, although overlapping, genealogies *and* conceptual decisions. Treating the history of biology in a more pluralistic manner could then contribute to widening the scope of philosophically pertinent problems.

Philosophy of biology before biology: biology and the conceptual conditions of its emergence

Various issues and challenges emerge here, most evidently the following: among all these programs and programmatic sketches mentioned above, which of them really corresponds to the emergence of 'biology' as such (a notion that actually needs to be pluralized, as we emphasize below)? And how is this science articulated or correlated with an ontology of life, that is, a type of inquiry that is concerned with the specific ontological status of living beings (without, however, being *grounded* on such considerations)? We suggest three conditions, three prerequisites for the emergence of biology – not analytic, but historical preconditions:

1 A 'phenomenal' dimension, corresponding to the perception of a new type of reality, populated by polyps, aphids, embryos, orangutans and monsters:[17] this is a kind of reality that is invisible to pure physics, or at least inexplicable in purely physical terms, as authors such as Maupertuis and Diderot, but also certain 'vitalists,' will emphasize.[18] One can also speak here of the 'contemplation' of a new reality, as Canguilhem does: 'A vitalist, I would suggest, is someone who is led to reflect on the nature of life more because of the contemplation of an egg than because she has handled a hoist or a bellows'; Diderot uses the same example and also stresses the 'seeing' of the egg, in more deliberately provocative terms: 'Do you see this egg? It is with this egg that we can overturn all schools of theology.'[19]
2 A more 'taxonomic' aspect, which reflects the reorganization and recomposition of natural history in the post-Buffon period, given that natural history is sometimes viewed as the very 'locus' of this reorganization of knowledge. This can be seen in Diderot's statement that a 'great revolution in the sciences' is coming, which he associates notably with 'the history of nature,' and in Maupertuis, who imagines what natural history might become if it is turned into a 'real science,' i.e., no longer just one treatise on animals among others, offering 'agreeable tableaus for us to contemplate'; instead, such a science should present 'the general processes of Nature, in its production and preservation.'[20] Such processes would amount to biological laws: natural history understood as a general theory of life – hence, a biology – would study 'the natural unity of living beings in terms of their organization and their functions.'[21] We find a similar situation in Kant's opposition between a mere 'description of Nature' (*Naturbeschreibung*), which he associates with Linnaeus, and the 'history of nature' (*Geschichte der Natur*), which he presents as new.[22]

   Linnaeus himself emphasized that classification into genera and species was the highest point that science (presumably life science) could reach

(Linnaeus 1751, §290). Similarly, a half-century later, in his 1802 treatise entitled *Biologie, oder Philosophie der lebenden Natur, für Naturforscher und Aerzte* ('Biology, or Philosophy of Living Nature, for Naturalists and Physicians'), Treviranus pointed out that, if botany and zoology are treated as parts of biology, they appear 'in a new light,' no longer mere chunks of nomenclature, because they take a step beyond traditional classifications: 'medicine, physiology and pathology will all be renewed'; Treviranus also quotes Stahl's insistence on how 'one first has to know what life is – what we commonly call "life."'[23]

3 Lastly, the appearance of biology is contingent upon the definition of the boundaries of a new science (whether this be termed 'biology,' 'zoonomy,' or otherwise), based on various domains such as morphology, embryology, or physiology. This is a 'criterial' aspect of the science, which can take different forms depending on the particular work and the context (on the question of 'criteria' for life, or living beings, see Malaterre 2010). Not only can these criteria vary, but they can be more or less strongly 'ontologized.' This is the case, quite differently, in Blumenbach, Lamarck, and Treviranus, with the focus on 'vital forces' or 'living bodies' as opposed to merely physical, brute, or dead matter.

Thus Blumenbach insists that his 'physiological' investigation deals with the 'vital forces' that 'belong exclusively to the organic matter of which we are made';[24] when Lamarck defines 'biology' (not so much in the unpublished manuscript bearing that name, as in his *Recherches sur l'organisation des corps vivants*, also from 1802), he does so with reference to 'living bodies':

> It includes everything pertaining to living bodies, especially their organization, the way it develops, its increasing composition throughout the prolonged activity of vital motions, its tendency to create specific organs, to isolate them, to centralize their action in a *foyer*, etc.[25]

Karl Friedrich Kielmeyer, who also sought to renew animal classification based on a comparative physiology of vital forces, also describes biology as a science of the 'animal kingdom.'[26] Sometimes this science is called 'physiology,' but there too we find the back-and-forth conceptual movement between the study of organic functions based on a combination of anatomy, vivisection, and, later, physico-chemical experimentation (Coleman 1977, 144), and the study of these functions *as vital* (as will still be the case in Claude Bernard). More broadly, Treviranus claimed to be inaugurating a science which was independent of medicine and natural history, the object of which was to study life in its various manifestations, and discover its laws. The introduction to his *Biologie* includes a chapter entitled 'Object and importance of biology,' in which he presents this new science as studying 'the various forms and manifestations of life,' its conditions and laws, and 'which causes govern its action.'[27]

Our question is not 'who' deserves to be at the center of the emergence of biology, or which of these projects is genuinely original, marking a 'discontinuity' or a 'break.'[28] Rather, we seek to locate and rethink this emergence in terms both of its conceptual conditions (what relations are there between the doctrines that immediately precede the constitution of biology as a science?) and its posterity (what relation is there between, e.g., the development of the organism concept and the transformations of biology?). In addition, these three conditions (1–3), if taken together, yield not 'biology' in the singular but 'biologies,' which highlights different possible dimensions (physico-chemical properties, organization, function, reproduction … which will be more or less unified in cell theory). This is a science or general theory of life – but if the nature of life itself is brought to the fore – as the crucial issue to be explained, to be investigated – then we are faced with the question of the relation between an *ontology* of life and a *science* of life.[29]

How could we determine if the constitution of biology responds, not just to internal movements of restructuring, as in the 'unification' of the natural history of living beings, but also to a kind of ontological challenge? Initially, the latter hypothesis may strike one as a bit quaint, overly 'internalist' in the sense this term has in the history of science, i.e., overly emphasizing ideas or the internal logic of key texts, rather than material practices and contexts. But this is indeed a rather different approach, one that typically relies on familiar schools of thought or positions such as mechanism versus vitalism, or 'how mechanism fails to grasp organism.'[30] What is missing here is the recognition of an independent conceptual dimension, as McLaughlin points to quite clearly – in his case, with respect to Kant:

> Kant is concerned not with the question of whether mechanism or vitalism (which arose in his lifetime) is right in biology but rather with the question of whether reductionism (which he considers to be the only scientific method) when applied to the organism displays a structural flaw that again and again necessitates teleological additions.
>
> (McLaughlin 1990, 3)

However, it is also the case that major actors in the story that concerns us here do express themselves in this way: for Maupertuis, 'the bodies of animals and of plants are machines that are too complicated' to be grasped in purely mechanical terms; 'we will never explain the formation of an organized body by the properties of matter alone.'[31] For Haller,

> In the animal, many machines are quite foreign to ordinary mechanical laws; tiny causes produce large movements; the speed of the humors is barely affected by causes which, according to the usual laws, should arrest it; violent movements and shortenings of fibres occur beyond any calculation, etc.[32]

But one can also have notions such as teleology or self-organization govern one's analysis of the emergence of biology (precisely, with an ontological dimension). This is typically the case of studies based on the Kantian and, especially, post-Kantian context, with *Naturphilosophie* (Lenoir 1981, 1982, Richards 2000, 2002, Zammito 2018). Certain histories of biology, just like some of the works of philosophical scholarship privileging the figure of Kant, are, in their entirety, histories of self-organization.[33] For indeed, it is often held that the characteristic properties of the systems studied by the life sciences are reducible to their self-organization. The capacity to reproduce and organize oneself is, classically, what makes the difference between us (but also butterflies and worms) and machines, such as ordinary watches. We are 'self-winding machines,' in La Mettrie's wonderful phrase.[34] The self-organizing character of living matter entails that 'the materials are the workers themselves' (Maupertuis).[35] In Kant, the notion of self-organization emerges in order to capture the 'relational' existence of the parts of an organism, which cannot exist separately, but do so rather in an interdependent way comprising an 'organized whole,' which is in turn a condition of the existence and functioning of these parts. It is this self-organizing feature of living systems that leads Kant to treat them as 'natural purposes.'[36] The existence of such properties will be treated as crucial in any distinction between living and nonliving beings, explicitly so in Lamarck.[37]

From the standpoint of biology's constitution as a science, the discovery and recognition of the existence of self-organizing mechanisms such as temperature regulation, metabolism – overall, everything pertaining to what will come to be called the 'robustness' of living systems (Piedrafita *et al.* 2010) – play an important role in the definition of the science. In Blumenbach, for instance, the study of biological functions is explicitly tied to a strong concept of vital force: his *Bildungstrieb* is truly a *Lebenskraft* that, alone, enables the explanation of how physical and chemical laws subserve the laws of organization in the course of embryogenesis.[38] Yet it is also possible to interpret this acknowledgment of self-organizing processes in living beings as, quite the contrary, an alternative to vital forces, and thus further removed from *Naturphilosophie*.

Indeed, rather than focusing on organisms as goal-oriented wholes, one could as well focus on the physico-chemical processes that are responsible for biological organization.[39] This, in turn, would allow many questions that have been relatively overlooked to be raised: first about the nature of this organization and its types (i.e., how to characterize plants, animals, etc.), second about the reasons for this organization, and then about its effects (i.e., to what extent is it responsible for the vital phenomena of interest?). Take, for instance, the points of contact between nascent biology and slightly more emerged chemistry – around what will come to be called organic chemistry.[40] If we examine these areas with an eye to the conceptual conditions of the emergence of biology, it then appears that one condition for this emergence – of biology as a science dedicated to the study of the specificity displayed by living beings – had less to do with the affirmation of irreducible vital forces, and lay more in a shift involving a new, chemically motivated conception of self-organization processes.[41] In this respect, we could emphasize the role played by a renewed and expanded account of nutrition in the late eighteenth century, in the emergence of the notion of self-organization

– how organisms build their own organization and maintain themselves – and the development of an epigenetic embryological framework in the constitution of biology, as exemplified by C. F. Wolff's renovated account of epigenesis and use of the *Nutritionskraft* and analogy between nutrition and generation in his renovated account of epigenesis (Wolff 1759, 1764, 1789). This shift in the understanding of nutrition, as an organizing process rather than a growth of a preexisting structure (as in Bonnet's use of the analogy between nutrition and embryonic development, e.g. Bonnet 1762, I, 3; I, 6), would therefore sustain the development of an epigenetic framework and influence the elaboration of a conception of organisms as self-organizing and self-sustaining entities, able to produce their organization and maintain their form despite their continuous material renewal.

Conclusion

We have taken the narrative of the genesis of biology away from the traditional focus on its 'founders,' without thereby handing away control over this narrative to those who monitor the lives of institutions, networks, or material practices. While from their standpoint it might seem as if we are turning the clock back in an internalist direction, with our evident conceptual focus, we would say to them that we have ceaselessly insisted on how historiographic and interpretive choices massively impact on the set of problems addressed in the philosophy of biology (from what it might myopically take to be a safe distance), given that these problems are directly shaped by their own tortuous *Begriffsgeschichte*. Conversely, to the defender of organicism in one corner or the promoter of neomechanism in another, both of whom regularly call for historical 'support' for their views, we suggest that our methodological provocation can run both ways.

Acknowledgments

We would like to thank Barnaby Hutchins for his careful reading of an earlier version of this text, and Peter McLaughlin for useful comments and suggestions.

Notes

1 Caron 1988, Barsanti 2000, McLaughlin 2002, Wolfe forthcoming.

2 Diderot's usage of the polyp as a metaphor is most developed in the 1769 *Rêve de D'Alembert*, but he refers to it also in the 1753 *Pensées sur l'Interprétation de la Nature*. Kant's specific essay on nervous disorders is his 1764 *Versuch über die Krankheiten des Kopfes*, but also his anthropology also vividly reflects this topic. As for Bonnet's Leibnizianism, he insists on it throughout his work (ironically, he was also accused by some, including Moses Mendelssohn, of having plagiarized Leibniz), and he viewed Trembley's discovery of the polyp as a confirmation of Leibniz's predictions concerning zoophytes (Marx 1976, 83). Bonnet explicitly sought to naturalize Leibniz's metaphysics, or at least to provide an empirical basis for it, by means of observation and analysis (Bonnet 1783, e.g., 15).

3 Treviranus' *Biologie* appeared in six volumes starting in 1802; Lamarck's *Hydrogéologie* in 1801–1802. In fact, Lamarck planned a treatise entitled *Biologie, ou Considérations sur la nature, les facultés, les développements et l'origine des corps vivants*, which he never completed. A part of the manuscript exists, but did not circulate; it was only published in the twentieth century by P.-P. Grassé (Lamarck 1944). Grassé considers the text to be from 1812–1814 but Pietro Corsi has shown more recently that *Biologie* represents older material, earlier than his 1802 book (Corsi 2006, 38n).

4 Roger, 'Le monde vivant' (1980), in Roger 1995, 210.

5 Roose 1797, Introduction (cf. Dittrich 1974); Schmid 1798–1801; Burdach 1800. On the history and development of biology in those years (1795–1802), see Barsanti 2000, Caron 1988, Kanz 2002, and Toepfer 2011.

6 McLaughlin 2002. Hanov uses the term in the text of volume 4 rather than 3, however. And again, 'biology' is not used in Hanov's text as we would understand it.

7 Schmid calls this science 'Zoonomia' (Schmid 1798, vol. I, 140, cit. Risse 1972, 153), as do Erasmus Darwin and others, as late as Royer-Collard in 1828; Kielmeyer lectures on 'allgemeine Zoologie' in 1806, describing this science as a 'physics of organized being' – in other words, a biology (Lenoir 1982, 50f.). Bernard defines physiology as the science of life (Bernard 1879, I, 1ère leçon, 3). Cf. Kanz 2002, 11.

8 de Richerand 1817, Prolégomènes, 1 (thanks to G. Toepfer for this reference); the first edition of Richerand's textbook appeared in 1801, but does not include this expression, even if it does include long discussions on the nature and scope of physiology, understood as the science that studies the entirety of properties of the animal economy, i.e., the science of life.

9 Gayon 2009, Pradeu 2017, and the reflections in Burian 1988 (as well as Nicholson and Gawne 2015 for a more revisionist view of the discipline, arguing for placing organicism at center stage); for early definitions of a 'biological philosophy' project somewhat closer to philosophical anthropology than to what was to become the philosophy of biology, Canguilhem 1947, 1968. As an interesting note, David Wiggins takes credit for calling for the emergence of the discipline, in the preface to his important work of 1967 *Identity and Spatio-Temporal Continuity*. Wiggins wrote about the centrality of biology to metaphysical investigation that 'It gradually became evident to me in constructing this work that for the future of metaphysics no single part of the philosophy of science was in more urgent need of development than the philosophy of biology' (Wiggins 1967: vii; see also Ferner 2016).

10 With reference to Whewell, one could then ask why there is no biology in Whewell's *History of the inductive sciences*, which came out in 1837, a year before Comte's volume 3 (1838). P. McLaughlin suggests (in discussion) that the term 'biology' comes into English with Whewell by way of Comte from Blainville and Treviranus.

11 Gayon 2009 and Pradeu 2017 have studied the distribution of the articles that appeared in the journal *Biology and Philosophy* (founded by Michael Ruse in 1986), which they treat as representative of work in philosophy of biology, between 1986 and 2015, according to which areas of biology were studied. Their analysis shows a clear predominance of evolutionary biology: 72 percent of published articles during the period studied by Gayon (1986–2002), and 62 percent for the period studied by Pradeu (2003–2015). An interesting contrast appears if one compares these results, as Pradeu does, with the division of areas in the Proceedings of the National Academy of Science (PNAS): here, biochemistry (12 percent), neuroscience (12 percent), medical science (10 percent), biophysics and computational biology (9 percent), immunology (7 percent) and microbiology (7 percent) predominate, while only 5 percent of the articles published in *PNAS* deal directly with evolutionary biology. Philosophy of biology then seems to focus primarily on one part of biology, in a disconnect with work in that science overall. Other philosophers have noted this disconnect as well:

> Traditionally, evolution has been the focus of most philosophical attention. While it surely remains true that 'nothing in biology makes sense except in light of evolution' … this tradition within the philosophy of biology is myopic insofar as it ignores much – if not most – of the work in contemporary biology.
>
> (Sarkar and Plutynski 2008, xviii)

12 For a conceptual and historical approach centered around the concept of organism, see e.g. Huneman and Wolfe 2010, and for a recent call for a revised historiography of biology, no longer centered around evolutionary biology, see Creager 2017. Interestingly, the term 'problem space' is also used by Lidgard and Nyhart in their recent, exhaustive essay on notions of biological individuality in the history of modern biology (Lidgard and Nyhart 2017, 24–25, 41–42): in their case it is borrowed from Newell and Simon, but they argue for a less formal (and mathematical), more historical and expanded problem space, and give an example similar to ours, of the way in which a fitness-centered problem space for the study of biological individuality will result in the exclusion of morphological or structural factors (which would then be a different problem space).

13 Gayon 2009, 205–206; Canguilhem 1980 [1968], 'Introduction. L'objet de l'histoire des sciences,' 11–12.

14 One important recent development that makes the above contrast less stark is the effort to renew connections between the history and the philosophy of science in the English-speaking world, with the society for Integrated History and Philosophy of Science (&HPS), which states on its website (and that of the 2nd &HPS conference, &HPS2), the following justification for such a connection:

> Good history and philosophy of science is not just history of science into which some philosophy of science may enter, or philosophy of science into which some history of science may enter. *It is work that is both historical and philosophical at the same time*. The founding insight of the modern discipline of HPS is that history and philosophy have a special affinity and one can effectively advance both simultaneously.
>
> (Emphasis ours; University of Notre Dame, 2009)

So far, such efforts have focused primarily on sciences such as physics and chemistry. Yet, with respect to biology proper, it seems obvious that philosophy of biology sought to take its distances from the more historical traditions in a way that was constitutive of its emergence as a professional field or subdiscipline: by definition, it excluded the historical analysis of concepts and the emergence of forms of knowledge, in favor of conceptual clarification and analysis, these being treated as equivalent to the work of the scientist (see Gayon 2009).

15 Gayon 1995, 464–465.

16 Duchesneau has greatly contributed to enrich this narrative, precisely by addressing the question of the emergence of biology through a focus on different subfields and their interplay with chemical sciences such as cell theory and general physiology: see e.g. Duchesneau 1987, 2004, 2010, and also Fox Keller 2010.

17 On polyps, Vartanian 1950; on orangutans, Smith 2007; on monsters, Wolfe ed. 2005.

18 We hope to show elsewhere how eighteenth-century vitalism and biology are interrelated in a deeper way than has been noticed thus far. There is some interesting terminological data ('organism,' 'vitalism' and 'biology' all stabilize as terms within the same decade, in the 1790s) but also, the focus on the 'animal economy' notion in vitalism (Wolfe and Terada 2008) is explicitly a program for studying laws and processes specific to living beings.

19 Canguilhem 1980, 88; Diderot, *Rêve de D'Alembert* (1769), in Diderot 1975, vol. XVII, 103–104.

20 Diderot, *Pensées sur l'interprétation de la nature* (1753/1754), §4, in Diderot 1975, vol. IX, 30–31; Maupertuis, *Lettre sur le progrès des sciences* (1752), §13, in Maupertuis 1756, vol. II, 386.
21 Canguilhem, 'Lamarck et Darwin' (1957), in Canguilhem 2015, 744.
22 Kant uses the language of *Naturbeschreibung* versus *Geschichte der Natur* in his writings on race including the 1775 *Von der Verschiedenheit der Racen Überhaupt*, in the essay on the use of teleological principles in philosophy (1788), and in the *Critique of the Power of Judgment*, §80 (thanks to P. Huneman for this point); in §79 he claims that his own account is descriptive. The distinction between natural history and description of nature was also important in Buffon's *Histoire Naturelle*. In Kant's essay on races (Ak. II, 429) and in the third *Critique* (Ak. V, 418–419), he also imagines that natural history could be reconfigured in terms of laws of relations between living individuals, studying the structural and functional similarities between members of a species, and their variations. For more on the distinction between 'description of nature' and 'history of nature,' see Sloan 2006 and Chapters 4 and 7 in Huneman 2007b.
23 'Ante omnia itaque scire convenit, quid sit illud quod vulgata appellatione vita dicitur' (Stahl, *Theoria Medica Vera*, 253, cit. Treviranus 1802, vol. I, 11n). Stéphane Schmitt has also noted that the anatomist Félix Vicq d'Azyr, who influenced Cuvier and played an important role in the structuring of biology as a discipline in nineteenth-century France, argues that anatomical and physiological inquiry needed to be tied to comparative study of animal functions (*Plan d'un Cours d'anatomie et de physiologie*, in Vicq d'Azyr 1805, vol. 4; Schmitt 2009).
24 Blumenbach 1787, part IV, 'De viribus vitalibus in universum,' §42, 32.
25 Lamarck 1802, 'Table raisonnée des matières,' 202. He also refers to biology with the curious term 'one of the three parts of terrestrial physics.'
26 See Zammito 2018, Chapter 9.
27 Treviranus 1802, vol. I, 4.
28 P. Corsi, and, in a less contextualized way, P. Charbonnat, put Lamarck in this position, because he considered life, before others did, as 'an organic phenomenon, the laws of formation of which result from own level of organization' (Charbonnat 2014, 226). Charbonnat considers this 'biology' to be a kind of third way, avoiding what he oddly portrays as a divide between vitalist physiology (Barthez) and materialist natural history (La Métherie, who is promoted to an unusually important role here). Others consider that Treviranus should occupy this position (cf. Lenoir 1981, Steigerwald 2014, Zammito 2018, 240–244), while earlier studies such as those of Schiller and Barsanti treated his project as unoriginal: Barsanti 1995, 210n). One can also reject this 'great man' vision of the history of biology, in favor of a 'collective' or 'network' vision of the emergence of this science, as is apparent in Corsi's large body of work on Lamarck, Cuvier, Geoffroy de Saint-Hilaire, and others (see e.g., Corsi 2011, Nyhart 2009). Thanks to P. Corsi for discussion on this point.
29 One of us has examined this relation (Wolfe 2011b and forthcoming).
30 For instances of such oppositions, see Gierer 1996 (on the animism–mechanism tension in the eighteenth century) and Allen 2005 (on the vitalism–mechanism tension in the early twentieth century). As for vitalism, recent historiography has depicted it as a more complex and protean entity, and has re-evaluated its impacts on the development of biology (Cimino and Duchesneau eds. 1997, Rey 2000, Williams 2003, Wolfe 2011a, 2017, Wolfe ed. 2008, Wolfe and Normandin eds. 2013, Steigerwald 2013 and the discussion in Gissis 2014).
31 *Essai de cosmologie*, 1750, in Maupertuis 1756, vol. I, 14; *Système de la nature*, §XXVIII, in op. cit., vol. II, 155–156.
32 Haller 1757, vol. I, livre I, Praefatio, v–vi (translation ours); cf. Duchesneau 1982, 127.

33 On the emergence of self-organization in the context of Kant and nascent biology, see Huneman 2008, and, for a less specifically philosophical treatment, see Sheehan and Wahrman 2015.
34 *L'Homme-machine*, in La Mettrie 1987, vol. I, 69.
35 Maupertuis, *Système*, §LXI, in Maupertuis 1756, vol. II, 180.
36 cf. Zammito 1992; Huneman 2007a, 2008; Steigerwald ed. 2006, 2010.
37 Fox Keller 2010, 9–10 and compare Gissis, Chapter 9 in this volume.
38 This is how Christoph Girtanner, among others, will summarize Blumenbach's idea. To be specific, Blumenbach distinguishes between five different types of vital force, of which the *Bildungstrieb* is the most general (these include nervous force or sensitivity, cellular force, muscular force or irritability, and the 'specific lives' which explain the movements of certain specific organs: Blumenbach 1787/1797, part IV, §§42–48; cf. Duchesneau 2011). For divergent interpretations of the relationship between Kant and Blumenbach, see Lenoir 1982, Richards 2000, Zammito 2012.
39 For a more organism-centered account of the 'philosophical history' of biology, see Grene and Depew 2004. For the chemical dimension that has not been studied so much: Antoine Lavoisier (1792, Lavoisier and Seguin 1789, 1790) for example contributed to initiate such a chemical investigation of the (self-) regulatory functions of the 'machine animale' (respiration, perspiration, nutrition). Later in the nineteenth century, Claude Bernard (1878), after his discovery of the glycogenic function of the liver (Bernard 1848, 1853), famously argued that life consisted in processes of organic creation (synthesis) and organic destruction (analysis), and that those processes were to be studied by a yet-unexisting 'physiological chemistry' that he called for.
40 On the history of organic chemistry and its interplay with biology, see Holmes 1963, 1974, 1992, Hall 1969, Needham 2008, Florkin 1972–1979, Fruton 1972, 1982, 2006, Kohler 1982, Klein 2003.
41 See for example Bognon-Küss, forthcoming. On the importance of a 'vital chemistry' at the intersection of science and philosophy as in the case of Diderot, see Pépin 2012.

References

Allen, G. E. 2005. Mechanism, vitalism and organicism in late nineteenth and twentieth-century biology: The importance of historical context. *Studies in History and Philosophy of Biological and Biomedical Sciences*, *36*: 261–283.

Barsanti, G. 1994. Lamarck and the birth of biology, 1740–1810, in S. Poggi and M. Bossi (eds.), *Romanticism in Science. Science in Europe 1790–1840* (pp. 47–74), Dordrecht: Springer.

Barsanti, G. 1995. La naissance de la biologie: Observations, théories, métaphysiques en France 1740–1810, in C. Blanckaert, J.-L. Fischer, and R. Rey (eds.), *Nature, Histoire, Société: essais en hommage à J. Roger* (pp. 197–227), Paris: Klincksieck.

Barsanti, G. 2000. Lamarck: Taxonomy and theoretical biology. *Asclepio*, *52*(2): 119–131.

Bernard, C. 1848. De l'origine du sucre dans l'économie animale. *Archives générales de médecine*, *18*: 303–319.

Bernard, C. 1853. *Nouvelle fonction du foie, considéré comme organe producteur de matière sucrée chez l'homme et les animaux*, Paris: J.-B. Baillière.

Bernard, C. 1878. *Leçons sur les phénomènes de la vie communs aux animaux et aux végétaux*, Paris: J.-B. Baillière et fils.

Bernard, C. 1879. *Leçons de physiologie opératoire*, Paris: J.-B. Baillière et fils.

Bichat, X. 1800. *Recherches physiologiques sur la vie et la mort*, Paris: Brosson, Gabon et Cie.

Blumenbach, J. F. 1787. *Institutiones physiologicae*, Göttingen: Dieterich. French edition 1797. *Institutions physiologiques*, trans. J.-F. Pugnet. Lyon: Reymann et Cie.

Bognon-Küss, C. Forthcoming. Between biology and chemistry in the eighteenth century: How nutrition shapes vital organization. *History and Philosophy of the Life Sciences*, special issue 'Organic, Organization, Organism. Essays in the history and philosophy of chemistry and biology.'

Bonnet, C. 1762. *Considérations sur les corps organisés*, Amsterdam: Marc Michel Rey.

Bonnet, C. 1764. *Contemplation de la nature*, Amsterdam: Marc Michel Rey.

Bonnet, C. 1769. *Palingénésie philosophique*, Geneva: C. Philibert.

Bonnet, C. 1783. Recueil de divers passages de Leibniz sur la survivance de l'animal, pour servir de Supplément à la partie VII de la *Palingénésie philosophique*, in C. Bonnet, *Œuvres d'histoire naturelle et de philosophie*, vol. 18: *Ecrits divers* (pp. 3–39), Neuchâtel: S. Faulche.

Burdach, K. F. 1800. *Propädeutik zum Studium der gesammten Heilkunst*, Leipzig: Breitkopf and Härtel.

Burian, R. M. 1988. A dialogue between philosophy and biology. *The Quarterly Review of Biology*, *63*(2): 193–195.

Canguilhem, G. 1947. Note sur la situation faite en France à la philosophie biologique. *Revue de métaphysique et de morale*, *52*(3/4): 322–332.

Canguilhem, G. 1968. Biologie et philosophie, in R. Klibansky (ed.), *Contemporary Philosophy, a Survey*, vol. II (pp. 387–394), Florence: La Nuova Italia.

Canguilhem, G. 1980. *La connaissance de la vie*, revised edition, Paris: J. Vrin.

Canguilhem, G. 2015. *Œuvres complètes, tome IV: Résistance, philosophie biologique et histoire des sciences (1940–1965)*, ed. C. Limoges, Paris: J. Vrin.

Caron, J. 1988. 'Biology' in the life sciences: A historiographical contribution. *History of Science*, *26*: 223–268.

Charbonnat, P. 2014. *Naissance de la biologie et matérialisme des Lumières*, Paris: Kimé.

Cheung, T. 2000. *Die Organisation des Lebendigen. Die Entstehung des biologischen Organismusbegriffs bei Cuvier, Leibniz und Kant*, Frankfurt: Campus-Verlag.

Cheung, T. 2006. From the organism of a body to the body of an organism: Occurrence and meaning of the word 'organism' from the seventeenth to the nineteenth centuries. *British Journal for the History of Science*, *39*(3): 319–339.

Cimino, G., and Duchesneau, F. (eds.). 1997. *Vitalisms from Haller to the Cell Theory*, Florence: Olschki.

Coleman, W. 1977. *Biology in the Nineteenth Century: Problems of Form, Function and Transformation*, Cambridge: Cambridge University Press.

Corsi, P. 2006. Biologie, in P. Corsi, J. Gayon *et al.* (eds.), *Lamarck, Philosophe de la nature* (pp. 37–64), Paris: PUF.

Corsi, P. 2011. The revolutions of evolution: Geoffroy and Lamarck, 1825–1840. *Bulletin du Musée d'Anthropologie préhistorique de Monaco*, *51*: 97–122.

Creager, A. N. H. 2017. A chemical reaction to the historiography of biology. *Ambix*, *64*(4): 343–359.

De Lamarck, J.-B. 1809. *Philosophie zoologique*, Paris: Dentu.

De Lamarck, J.-B. 1944. *Biologie ou Considérations sur la nature, les facultés, les développemens et l'origine des corps vivans* (1800), ed. P.-P. Grassé, *La Revue Scientifique*, *82*: 267–276.

Diderot, D. 1975. *Œuvres complètes*, eds. H. Dieckmann, J. Proust, and J. Varloot, Paris: Hermann.

Dittrich, M. 1974. Progressive Elemente in den Lebensdefinitionen des romantischen Naturphilosophie. *Communicationes de historia artis medicinae*, *78–79*: 72–85.

Duchesneau, F. 1982. *La physiologie des Lumières. Empirisme, modèles, theories*, The Hague, London, Boston: M. Nijhoff.

Duchesneau, F. 1987. *Genèse de la théorie cellulaire*, Montréal, Paris: Bellarmin, Vrin.

Duchesneau, F. 2004. L'organisation du vivant: Émergence ou survenance? *Sens Public*. http://sens-public.org/article856.html.

Duchesneau, F. 2010. Rôle du couple 'structure/fonction' dans la constitution de la biologie comme science, in J. Gayon and A. de Ricqlès (eds.), *Les fonctions: des organismes aux artefacts* (pp. 43–64), Paris: PUF.

Duchesneau, F. 2011. Blumenbach et la théorie des forces vitales, in P. Nouvel (ed.), *Repenser le vitalisme* (pp. 73–88), Paris: PUF.

Duchesneau, F. 2018. *Organisme et corps organique de Leibniz à Kant*, Paris: Vrin.

Ferner, A. M. 2016. *Organisms and Personal Identity: Biological Individuation and the Work of David Wiggins*, London: Routledge.

Florkin, M. 1972–1979. *A History of Biochemistry*, Amsterdam, London, New York: Elsevier.

Fruton, J. 1972. *Molecules and Life; Historical Essays on the Interplay of Chemistry and Biology*. New York: Wiley-Interscience.

Fruton, J. 1982. *A Bio-Bibliography for the History of the Biochemical Sciences since 1800*, Philadelphia, PA: American Philosophical Society.

Fruton, J. 2006. *Fermentation: Vital or Chemical Process?* Leiden, Boston: Brill.

Fox Keller, E. 2010. Self-organization, self-assembly, and the inherent activity of matter, the Hans Rausing lecture 2009 (Uppsala University, Department of History of Science and Ideas). *Salvia Småskrifter*, *12*: 1–26.

Gayon, J. 1995. La philosophie biologique dans l'œuvre historique de Jacques Roger. Postscript to J. Roger, *Pour une histoire des sciences à part entière*, ed. C. Blanckaert (pp. 459–471), Paris: Albin Michel.

Gayon, J. 2009. Philosophy of biology: An historico-critical characterization, in A. Brenner and J. Gayon (eds.), *French Studies in the Philosophy of Science: Contemporary Research in France* (pp. 201–212), Dordrecht: Springer.

Gierer, A. 1996. Organisms-mechanisms: Stahl, Wolff and the case against reductionist exclusion. *Science in Context*, *9*(4): 511–528.

Gissis, S. B. 2014. The continuing vitality of the *problématique* of vitalism? *Studies in History and Philosophy of Biological and Biomedical Sciences*, *47*: 196–200.

Goldstein, K. 1934. *Der Aufbau des Organismus: Einführung in die Biologie unter besonderer Berücksichtigung der Erfahrungen am kranken Menschen*, The Hague: Martinus Nijhoff.

Grene, M., and Depew, D. 2004. *The Philosophy of Biology: An Episodic History*, Cambridge: Cambridge University Press.

Hall, T. S. 1969. *Ideas of Life and Matter. Studies in the History of General Physiology*, Chicago, IL: University of Chicago Press.

Haller, A. von. 1757. *Elementa physiologiæ corporis humani*, vol. I, Lausanne: Marc-Michel Bousquet.

Hanov, M. C., 1766. *Philosophiae naturalis sive physicae dogmaticae tomus III, continens geologiam, biologiam, phytologiam generalem et dendrologiam vel terrae, rerum viventium et vegetantium in genere, atque arborum scientiam*, Halle: n.p.

Holmes, F. L. 1963. Analysis and the origins of physiological chemistry. *Isis*, *51*(1): 50–81.

Holmes, F. L. 1974. *Claude Bernard and Animal Chemistry: The Emergence of a Scientist*, Cambridge, MA: Harvard University Press.

Holmes, F. L. 1992. *Between Biology and Medicine: The Formation of Intermediary Metabolism*, Berkeley, CA: Office for History of Science and Technology, University of California at Berkeley.

Hull, D. 1969. What philosophy of biology is not. *Journal of the History of Biology*, *2*: 241–268.

Huneman, P. 2006. Naturalising purpose: From comparative anatomy to the 'adventure of reason.' *Studies in History and Philosophy of Science Part C: Studies in History and Philosophy of Biological and Biomedical Sciences*, *37*(4): 649–674.

Huneman, P. 2007a. Reflexive judgment and Wolffian embryology: Kant's shift between the first and the third *Critiques*, in P. Huneman (ed.), *Understanding Purpose, Kant and the Philosophy and Biology* (pp. 75–100), Rochester, NY: University of Rochester Press.

Huneman, P. (ed.). 2007b. *Understanding Purpose, Kant and the Philosophy and Biology*, Rochester, NY: University of Rochester Press/North American Kant Society Studies in Philosophy.

Huneman, P. 2008. *Métaphysique et biologie. Kant et la constitution du concept d'organisme*, Paris: Kimé.

Huneman, P., and Wolfe, C. T. 2010. Introduction. *History and Philosophy of the Life Sciences, 32*(2/3), Special issue, 'The Concept of Organism: Historical Philosophical, Scientific Perspectives.'

Jonas, H. 1966. *The Phenomenon of Life. Towards a Philosophical Biology*, New York: Harper & Row.

Kant, I. 1796. *Über das Organ der Seele*, trans. as *From Soemmerring's On the Organ of The Soul*, in I. Kant, *Anthropology, History, and Education* (AA 12, 31–35), eds. G. Zöller and R. B. Louden, Cambridge: Cambridge University Press, 2007.

Kanz, K. T. 2002. Von der BIOLOGIA zur Biologie. Zur Begriffsentwicklung und Disziplingenese vom 17. bis zum 20. Jahrhundert, in U. Hoßfeld and T. Junker (eds.), *Die Entstehung biologischer Disziplinen II. Beiträge zur 10. Jahrestagung der DGGTB in Berlin 2001* (Verhandlungen zur Geschichte und Theorie der Biologie, 9) (pp. 9–30), Berlin: VWB.

Klein, U. 2003. *Experiments, Models, Paper Tools: Cultures of Organic Chemistry in the Nineteenth Century*, Stanford, CA: Stanford University Press.

Kohler, R. E. 1982. *From Medical Chemistry to Biochemistry: The Making of a Biomedical Discipline*, Cambridge: Cambridge University Press.

La Mettrie, J. O. de. 1987. *Œuvres Philosophiques*, ed. F. Markovits, 2 vols., Paris: Fayard.

Lavoisier, A. L. 1792. Prix proposé par l'Académie des Sciences pour l'année 1794. *Mémoires de l'Académie des sciences.*

Lavoisier, A. L., and Seguin, A. J. F. 1789. Premier mémoire sur la respiration des animaux. *Mémoires de l'Académie des sciences*, 185.

Lavoisier, A. L., and Seguin, A. J. F. 1790. Premier mémoire sur la transpiration des animaux. *Mémoires de l'Académie des sciences*, 77.

Lenoir, T. 1981. The Göttingen School and the development of transcendental Naturphilosophie in the Romantic era. *Studies in History of Biology*, *5*: 111–205.

Lenoir, T. 1982. *The Strategy of Life. Teleology and Mechanics in Nineteenth-Century German Biology*, Chicago, IL: University of Chicago Press.

Lidgard, S., and Nyhart, L. K. 2017. The work of biological individuality: Concepts and contexts, in S. Lidgard and L. K. Nyhart (eds.), *Biological Individuality: Integrating*

Scientific, Philosophical, and Historical Perspectives (pp. 17–52), Chicago, IL: University of Chicago Press.

Linnaeus, C. 1751. *Philosophia botanica*, Stockholm: Godofr. Kiesewetter.

McLaughlin, P. 1990. *Kant's Critique of Teleology in Biological Explanation: Antinomy and Teleology*, Lewistown, NY: Edwin Mellen Press.

McLaughlin, P. 2002. Naming biology. *Journal of the History of Biology*, *35*: 1–4.

Malaterre, C. 2010. On what it is to fly can tell us something about what it is to live. *Origins of Life and Evolution of the Biosphere*, *40*(2): 169–177.

Marx, J. 1976. *Charles Bonnet contre les Lumières, 1738–1750* (*Studies on Voltaire and the Eighteenth Century*, vols. 156–157), Oxford: Voltaire Foundation.

Maupertuis, P.-L. M. de. 1756. *Œuvres*, 2 vols., Lyon: Bruyset.

Meiners, C. 1800. *Allgemeine kritische Geschichte der Älteren und neueren Ethik oder Lebenswissenschaft nebst einer Untersuchung der Fragen: Gibt es dann auch wirklich eine Wissenschaft des Lebens?*, vol. 1, Göttingen: J. C. Dieterich.

Needham, J. 2008. *The Chemistry of Life: Eight Lectures on the History of Biochemistry*, Cambridge: Cambridge University Press.

Nicholson, D., and Gawne, R. 2015. Neither logical empiricism nor vitalism, but organicism: What the philosophy of biology was. *History and Philosophy of the Life Sciences*, *37*(4): 345–381.

Nyhart, L. K. 2009. *Modern Nature. The Rise of the Biological Perspective in Germany*, Chicago, IL: University of Chicago Press.

Pépin, F. 2012. *La philosophie expérimentale de Diderot et la chimie*, Paris: Garnier.

Piedrafita, G., Montero, F., Moran, F., Cardenas, M. L., and Cornish-Bowden, A. 2010. A simple self-maintaining metabolic system: Robustness, autocatalysis, bistability. *PLoS Computational Biology*, *6*(8): e1000872.

Pradeu, T. 2017. Thirty years of *Biology & Philosophy*: Philosophy of which biology? *Biology and Philosophy*, *32*(2): 149–167.

Rey, R. 2000. *Naissance et développement du vitalisme en France de la seconde moitié du 18e siècle à la fin du Premier Empire*, Oxford: Voltaire Foundation.

Richards, R. J. 2000. Kant and Blumenbach on the Bildungstrieb: A historical misunderstanding. *Studies in the History and Philosophy of Biology and Biomedical Sciences*, *31*: 11–32.

Richards, R. J. 2002. *The Romantic Conception of Life: Science and Philosophy in the Age of Goethe*, Chicago, IL: University of Chicago Press.

Richerand, A. de. 1817. *Nouveaux Éléments de physiologie*, 2 vols., 7th edition, Paris: Caille et Ravier.

Risse, G. 1972. Kant, Schelling, and the early search for a philosophical 'science' of medicine in Germany. *Journal of the History of Medicine*, *27*: 145–158.

Roger, J. 1963. *Les sciences de la vie dans la pensée française du XVIIIe siècle. La génération des animaux de Descartes à l'Encyclopédie*, Paris: Armand Colin.

Roger, J. 1995. *Pour une histoire des sciences à part entière*, ed. C. Blanckaert, Paris: Albin Michel.

Roose, T. G. A. 1797. *Grundzüge der Lehre von der Lebenskraft*, Braunschweig: Christian Friedrich Thomas.

Sarkar, S., and Plutynski, A. (eds.). 2008. *A Companion to the Philosophy of Biology*. Malden, MA: Blackwell.

Schmitt, S. 2009. From physiology to classification: Comparative anatomy and Vicq d'Azyr's plan of reform for life sciences and medicine (1774–1794). *Science in Context*, *22*(2): 145–193.

Sheehan, J., and Wahrman, D. 2015. *Invisible Hands. Self-Organization and the Eighteenth Century*, Chicago, IL: University of Chicago Press.

Sloan, P. 2003. Whewell's philosophy of discovery and the archetype of the vertebrate skeleton: The role of German philosophy of science in Richard Owen's biology. *Annals of Science*, *60*: 39–61.

Sloan, P. 2006. Kant on the history of nature: The ambiguous heritage of the critical philosophy for natural history. *Studies in the History and Philosophy of Biological and Biomedical Sciences*, *37*(4): 627–648.

Smith, J. E. H. 2007. Language, bipedalism and the mind–body problem in Edward Tyson's *Orang Outang* (1699). *Intellectual History Review*, *17*(3): 291–304.

Sömmering, S. T. 1796. *Über das Organ der Seele*, Königsberg: Nicolovius.

Steigerwald, J. (ed.). 2006. *Kantian Teleology and the Biological Sciences*, Special Issue of *Studies in History and Philosophy of the Biological and Biomedical Sciences*, *37*(4).

Steigerwald, J. 2010. Natural purposes and the reflecting power of judgment: The problem of the organism in Kant's critical philosophy. *European Romantic Review*, *21*(3): 291–308.

Steigerwald, J. 2013. Rethinking organic vitality in Germany at the turn of the nineteenth century, in S. Normandin and C. T. Wolfe (eds.), *Vitalism and the Scientific Image in Post-Enlightenment Life Science, 1800–2010* (pp. 51–76), Dordrecht: Springer.

Steigerwald, J. 2014. Treviranus' biology: Generation, degeneration and the boundaries of life, in S. Lettow (ed.), *Gender, Race, and Reproduction. Philosophy and the Early Life Sciences in Context* (pp. 105–127), Albany, NY: SUNY Press.

Treviranus, G. R. 1802–1822. *Biologie oder die Philosophie der lebenden Natur*, Göttingen: J. F. Röwer.

University of Notre Dame. 2009. *&HPS2*, 2nd Integrated History and Philosophy of Science Conference, March 12–15, 2009. Conference statement at https://reilly.nd.edu/news-and-events/conferences/archives/hps and at the Society for Integrated History and Philosophy of Science's webpage: http://integratedhps.org/en/about.

Vartanian, A. 1950. Trembley's polype, La Mettrie, and 18th century French materialism. *Journal of the History of Ideas*, *11*(3): 259–286.

Wiggins, D. 1967. *Identity and Spatio-Temporal Continuity*, Oxford: Blackwell.

Williams, E. 2003. *A Cultural History of Medical Vitalism in Enlightenment Montpellier*, Burlington, VT: Ashgate.

Wolfe, C. T. (ed.). 2005. *Monsters and Philosophy*, London: King's College Publications.

Wolfe, C. T. (ed.). 2008. *Vitalism without Metaphysics? Medical Vitalism in the Enlightenment*, special issue of *Science in Context*, *21*(4).

Wolfe, C. T. 2010. Do organisms have an ontological status? *History and Philosophy of the Life Sciences*, *32*(2–3): 195–232.

Wolfe, C. T. 2011a. From substantival to functional vitalism and beyond: Animas, organisms and attitudes. *Eidos*, *14*: 212–235.

Wolfe, C. T. 2011b. Why was there no controversy over life in the Scientific Revolution?, in V. Boantza and M. Dascal (eds.), *Controversies in the Scientific Revolution* (pp. 187–219). Amsterdam: John Benjamins.

Wolfe, C. T. 2017. Models of organic organization in Montpellier vitalism. *Early Science and Medicine*, *22*: 229–252.

Wolfe, C. T. Forthcoming. *La philosophie de la biologie avant la biologie: une histoire du vitalisme*, Paris: Garnier.

Wolfe, C. T., and Normandin, S. (eds.). 2013. *Vitalism and the Scientific Image in Post-Enlightenment Life Science, 1800–2010*, Dordrecht: Springer.

Wolfe, C. T., and Terada, M. 2008. The animal economy as object and program in Montpellier vitalism. *Science in Context*, *21*(4): 537–579.

Wolff, C. F. 1759. *Theoria Generationis*, Halle: Hendel. Rpt., Hildesheim: G. Olms, 1966.

Wolff, C. F. 1764. *Theorie von Der Generation in zwo Abhandlungen erklärt und bewiesen*. Berlin: Friedrich Wilhelm Birnstiel. Rpt., Hildesheim: G. Olms, 1966.

Wolff, C. F. 1789. Von der eigenthümlichen und wesentlichen Kraft der vegetabilischen sowohl als auch der animalischen Substanz. In *Zwo Abhandlungen über die Nutritionskraft ... Nebst einer fernen Erläuterung eben Derselben Materie von C. F. Wolff*. St. Petersburg: Kayserl. Akademie der Wißenschaften.

Woodger, J. H. 1929. *Biological Principles: A Critical Study*. New York: Harcourt, Brace.

Zammito, J. H. 1992. *The Genesis of Kant's Critique of Judgment*, Chicago, IL: University of Chicago Press.

Zammito, J. H. 2012. The Lenoir thesis revisited: Blumenbach and Kant. *Studies in History and Philosophy of Biological and Biomedical Sciences*, *43*: 120–132.

Zammito, J. H. 2018. *The Gestation of German Biology. Philosophy and Physiology from Stahl to Schelling*, Chicago, IL: University of Chicago Press.

Part I

Form and development

2 Buffon's theories of generation and the changing dialectics of molds and molecules

Stéphane Schmitt

Introduction

An important cause of the rise of the "science of life," that is to say, of the notion of life as an object of science around 1800, was the reflections on the ontogenesis or organisms that took place throughout the eighteenth century, and, more specifically, the emerging idea that the phenomena of generation established a clear borderline between living and nonliving bodies. Since the late seventeenth century, the triumph of mechanism had led to the common idea (even though resistance did exist) that all vital functions of animals and plants could be simply explained by the mere laws of general physics. As regards reproduction and development, for instance, the prevailing theories from the 1680s onwards dodged the difficult problem of ontogenesis and saw the apparent formation of new organisms at each generation as a mere (mechanical) growth of preexisting germs, created at the origin of the world. As a consequence, when a series of attacks were launched against preexistence in the mid-eighteenth century, they contributed to raise debates around the nature and specificity of life, the existence of "vital forces" and their exact nature, and the properties of the elementary components of the living beings, as well as the origin of their organization (in the course of the development or in the history of the Earth). Caspar Friedrich Wolff's studies on epigenesis, for instance, were particularly significant in that respect (see Zammito's chapter in this volume).

In this chapter, I examine the contribution of the French naturalist Georges-Louis Leclerc Buffon to these controversies. As early as 1749, Buffon proposed a new theory of generation and introduced two concepts: on the one hand, "organic molecules," the components of the organic matter, endowed with particular properties; and, on the other hand, "interior molds" that supposedly organized the molecules into organisms in the course of the embryonic development. The association of these two elements was, for Buffon, what characterized the living species and the perpetuation of a constant form from one generation to the other. His views were widely known from 1749 onwards and, even though they were mostly not accepted as such, they had a lot of influence in France as well as in other countries (especially in Germany).

Much attention has been put to Buffon's conception of generation by historians. However, here we argue that not enough stress has been laid on the existence, in his writings, of different views on the dialectics of molecules and molds, and that the part of morphogenetic information he acknowledged to each of them significantly varied in the course of his career. There were, as it were, several theories of generation in Buffon's work from the 1730s to the 1770s, each of them giving more or less important organizing powers to the molecules, and precising more or less the physical nature of these powers. These differences had major consequences on the conception of life and its specificity.

Generation before Buffon: the first attacks against preexisting germs by Bourguet and Maupertuis

Every theory of generation, from Aristotle to molecular biology, has had to make sense of both the constancy of reproduction (cats produce cats, dogs produce dogs, and children resemble their parents more closely than other people) and the plasticity of development (brothers and sisters are not exactly alike, hybrids or malformations sometimes occur, etc.). For about two millennia, Aristotle's and Galen's solutions to this problem and their avatars were satisfactory, in connection to the ancient physics that allowed the existence of various informational nonmaterial entities (Aristotelian form, souls, etc.) capable of organizing gradually the unorganized seminal matter into an embryo, i.e., fulfilling epigenesis. But when this kind of physics collapsed, in the course of the seventeenth century, such entities were banned and epigenesis waned too.[1] William Harvey (1651) certainly attempted to revive Aristotelian epigenesis, but, despite the quality of his observations, he did not succeed in reversing the trend.[2] On the other hand, René Descartes tried to make the new mechanistic paradigm he was building himself compatible with an epigenetic conception of embryogenesis, and he imagined that the particles of the seeds were able to self-organize into an embryo according to the laws of simple mechanics alone (see Aucante 2006). But the Cartesian theory of generation was not successful either.

Thus new solutions had to be found that could fit the new context. The notion of preexistence of germs emerged then. According to this, all germs, for all organisms, had been created (by God) once for all, at the beginning of the world, and had just to unfold and grow at each generation. Several forms of this theory appeared simultaneously in the last three decades of the seventeenth century, in connection with the recent discovery of spermatozoa ("animalcules") and eggs of viviparous animals. The most widespread view, called ovism, claimed that the preexisting germs were encapsulated in the eggs of the females, whereas a few followers of animalculism preferred to locate germs in spermatic animalcules. In both cases, the germs were supposed to be encased one inside the other, like Russian dolls, so that the smallness of the most internal ones was almost inconceivable. However, these theories seemed to be acceptable in the mechanist frame and preferable to the notion of *de novo* ontogenesis of organisms at each generation (Roger 1993).

Nevertheless, the preexistence of germs encountered a number of considerable difficulties. As a matter of fact, it made sense of the constancy in generation, but was unable to explain plasticity, i.e., the fact that one individual is not the exact copy of one single parent. In particular, double (paternal and maternal) heredity remained obscure, especially in the case of interspecific hybrids. Similarly, the birth of abnormal children or animals, or the regeneration of organs in certain organisms (e.g., Abraham Trembley's polyp) were inexplicable. Ad hoc solutions to these problems were proposed, such as maternal imagination, which could account for double heredity and monsters. Such arguments made preexistence credible for many decades, up to the mid-eighteenth century and beyond, but they could not prevent attacks, especially after the 1740s.

As early as 1729, Louis Bourguet (1678–1742), a Swiss naturalist and a correspondent of Leibniz, although a follower of ovist preexistence, emphasized the limits of this theory, for example its inability to explain paternal and maternal heredity, and he attempted to improve it by introducing the notion of the mold.[3] He acknowledged the preexistence of a germ in the maternal egg, but, rather than a small, perfect animal, he considered it a kind of mold into which the particles of the parental semens could insert themselves. He acknowledged the existence of a female semen, which was, just like the male semen, an "extract of the parts of the animal" (1729: 149). He thought that the mixture of both semens after copulation provided the germ with its first particles, which were, thus, already prepared by the body of both parents. Each particle could enter the preexisting germ (by "intussusception"), go in the right place, and participate in the formation and growth of the corresponding organ of the embryo.

This theory explained several phenomena such as maternal and paternal heredity and the existence of abnormalities (1729: 93–95). More generally, it introduced in the notion of preexisting germs a certain amount of plasticity, and conferred to living matter a certain activity. This broke away from the passivity of matter as understood by the (Cartesian) mechanists of the late seventeenth century such as Malebranche (see, among others, Roger 1993 [1963]). For Bourguet, each of the corpuscles that made up the bodies was "endowed with a vital activity corresponding to its shape" (66). He explained all vital phenomena, and more specifically the intussusception of the molecules into the mold, by the Leibnizian notion of "organic mechanism," that is, the monad as an immaterial unit that permitted the unity and the working of the organism (164–165).

The relation of the mold to the molecules was clear in Bourguet's system, since it was a mere variant of the preexistence theory: the molecules could be inserted into the organism only because this organism was already present materially as a germ (even if this germ was an incomplete frame) and because it had in itself the principle that made possible their insertion at the right place. The vital activity of the molecules was not strong enough for them to organize themselves autonomously: they were subordinate to an "activity" of a higher level (the monad) that organized them. In that respect, Bourguet kept quite close to Leibniz's thought (see Smith 2011: 155–156). However, the idea that the molecules that made up the living beings were material entities endowed with a certain vital activity, that they had a certain specificity and were associated with monads of lower levels, was not so Leibnizian. This suggested the existence of a hierarchy of monads, the most complex corresponding to higher organisms, the simplest to elementary corpuscles. Thus Bourguet opened the way to the representation of the living organism as a hierarchy of morphological as well as physiological individualities. He represented, as it were, a generational intermediary between, on the one hand, Leibniz, Newton, and the first proponents of preexistence (De Graaf, Malebranche, etc.), and, on the other hand, the new generations that developed a radical criticism of preexistence after 1740.

In the *Vénus Physique* (1745), Maupertuis listed the evidence against preexistence, he praised Harvey's forgotten observations, and he proposed a new epigenetic theory relying on a Newtonian model.[4] According to him, development took place in a mixture of male and female semens: the particles attracted or repelled each other and organized themselves into an embryo, as the result of forces analogous to Newtonian gravitation. In this model, most of the information was to be found in the forces and interactions between the particles rather than within the particles themselves.

However, at the end of the *Vénus Physique*, Maupertuis sketched a somewhat different theory, and he asked if the particles did not possess a kind of instinct, or will, that would enable their spontaneous self-organization. This idea can be seen as the result of the influence of a freely interpreted Leibnizianism on Maupertuis's thought, since, in this new model, the "parts of the semens" (that is, the smallest units of living matter) were endowed with a specific teleological activity (see Wolfe 2010).

This alternate version of his theory was much more developed a few years later (Maupertuis 1768 [1751]). Then Maupertuis minimized the specificity of the attractive forces in the formation of the living beings, and he did not believe any longer that such forces were able to sort out the particles into a complex structure like an organism. He clearly set the organizational specificity within the particles themselves rather than in the network of their interactions, and he felt obliged to assume that the particles had "a principle of intelligence, something like what we call *desire, aversion, memory*" (146). In this model, each particle in the mixture of semens recalled its previous position in the paternal or maternal bodies and took this position again in the fetus (158–159). In this way, Maupertuis was able to explain all the normal and abnormal phenomena of reproduction and heredity.

By providing the particles with intelligence, perception, and memory, that is, a kind of autonomy, Maupertuis opened the way to a wide range of new attempts to locate any physical or vital principle or property in these particles in order to make sense of their self-organization into an organism; depending on the nature of this principle, this could lead to a sort of vitalism, or to radical forms of materialism. Maupertuis himself was not specific on this point; he was probably influenced, at that point, by a kind of Leibnizian mechanism as he understood it. Anyway, he endowed living matter with powers that would have been unthought-of in a Cartesian or Newtonian context, and he viewed these powers at the level of the particles that made up organisms rather than at a higher level.

Furthermore, since his theory made spontaneous generation possible, Maupertuis imagined that, on the primitive Earth, all living beings may have appeared as the consequence of combination of particles, that these beings may have been gradually modified from one generation to another, and that all species may have arisen from one common ancestor (164–165). Interestingly, thus, the emergence of transformism in Maupertuis's thought is closely connected with the revival of epigenesis.

Buffon's first views on generation (1730s)

Buffon began his reflections on generation as soon as the early 1730s.[5] At that time he exchanged letters with Bourguet through his friend Bouhier, but he put an end to the discussion because of their divergences (Buffon 1971, 1: 40; see also Hanks 1966: 70–71). He was probably inspired to some extent by Maupertuis, who was his friend, but, even though he mentioned the *Vénus Physique* in 1749, he did not give more details on this influence (Buffon 1749–1767, 2: 164).

Indeed, Buffon's first views on generation are poorly known, except for what he wrote in later texts:

> [B]efore I had carefully examined the question of generation, I was taken with certain ideas of a mixed system where I used the spermatic worms and the eggs of the females as the first organic parts forming the living point, to which I supposed, like Harvey, the others joined themselves in a symmetrical and relative order through forces of attraction.
>
> (1749–1767, 2: 68–69)

This first system was thus impregnated with the notion of preexisting germs, even if the initial germs (or "living points"), coming from either the spermatic animalcules or the eggs (or maybe resulting from the union of two "half-germs" coming, respectively, by an animalcule and an egg: Buffon is not clear on that point), were seemingly incomplete and provided only the central, asymmetric parts of the body. The embryo had then to be completed with a gradual addition of parts, in an epigenetic way, by the means of Newtonian-like forces. This second step was similar both to Maupertuis's theory and to Buffon's model proposed in the last chapters of 1749 (see below), which suggests that Buffon retained something of his first ideas on generation in later writings. But he was not satisfied with them as a whole, since they left unexplained the "resemblances" of the children to their parents, and he turned to another theory.

Buffon's main theory of generation in 1749: organic molecules and interior molds

Buffon's most famous theory of generation was published in 1749, as the first part of the second volume of his *Histoire Naturelle, Générale et Particulière*, under the title "Histoire générale des animaux," which comprised 11 chapters

(Buffon 1749–1767, 2: 1–426). Two parts can be distinguished in this text, corresponding to two different aspects of the theory that are complementary to some extent but also, surprisingly, rather independent from each other. Let us consider the first nine chapters at first.

First of all, Buffon emphasizes the most essential property of living beings, both animals and plants, which radically distinguishes them from the inorganic world, namely the faculty of producing one's likeness. Instead of "generation," he deliberately uses the term "reproduction," which, in the eighteenth century, more commonly referred to the regeneration of organs. By doing that, he claims that the process is a formation of a new organism, and not the mere growth of a preexisting germ.

For him, this ability to reproduce determines the main gap in nature between, on the one hand, animals and plants and, on the other hand, minerals. As a consequence, the notion of species, according to Buffon, concerns only the former, since he defines a species as a group of individuals capable of breeding and producing a fertile offspring (Buffon 1749–1767, 2: 10–11). This criterion of interfertility makes an experimental approach of species possible and suggests a research program specific to living beings. Indeed, this question is omnipresent in the *Histoire Naturelle*, and Buffon tries to breed a number of animals (e.g., dogs with foxes or wolves) to determine if they belong to the same species.[6] Furthermore, this definition establishes a close connection between the problem of species and the theory of generation.

In order to explain the mechanism of reproduction, Buffon starts from the study of simple creatures, such as plants or polyps, which are able to reproduce a whole organism from small fragments. By analogy with the structure of crystals of salt, which are cubes made up of an infinity of smaller cubes, he infers that the living beings are made up of a considerable number of similar "primary and constituent parts" (Buffon 1749–1767, 2: 19–20). He uses several phrases to refer to these elements, such as "primitive and incorruptible parts" and "living organic parts" (Buffon 1749–1767, 2: 24, 40), but his favorite expression is "organic molecules" (Buffon 1749–1767, 2: 49).

He considers that the organic molecules are the same in the whole organic world, in plants as well as animals. They are particularly abundant in certain matters with a powerful activity, such as snake venom and, above all, the seminal matters (seeds of plants and seminal liquids of animals). The generation of every living being is nothing but the well-ordered assembling of organic molecules coming from the parents' semens. Conversely, the molecules disperse after death, following an eternal cycle without alteration (Buffon 1749–1767, 2: 44).

The exact nature of the organic molecules in Buffon's thought is rather vague. Their incorruptible nature is reminiscent of Lucretian atoms, while their "living" character has something in common with the particles of Maupertuis's second theory (see above). But the meaning of "living" is not clear, which, more generally, raises the question of Buffon's position toward the vitalist trends flourishing from the mid-eighteenth century (see Rey 1992; 2000; Wolfe and Terada 2008). He emphasizes the specificity of life in comparison to inanimate matter, but it is difficult to know if this specificity remains in the framework of Newtonian mechanism or if it results from the existence of a vital principle, irreducible to the general laws of physics and even to Newtonian-like forces. Some sentences strongly suggest a materialistic interpretation of life: for instance, Buffon sees life as a general trend in Nature, which "seems to strive to life much more than to death" (Buffon 1749–1767, 2: 37); he adds that "the organic is the most common product of nature, apparently the one which costs it the less" (Buffon 1749–1767, 2: 39). He even states that "the living and the animated [*le vivant et l'animé*], instead of being a metaphysical degree of beings, is a physical property of matter" (Buffon 1749–1767, 2: 17). This point is more explicit in later texts (see the last section).

Anyway, "living" is not a synonym of "organized" in Buffon's view, and he prefers to establish a distinction between living and dead matter, rather than between organized and brute matter (Buffon 1749–1767, 2: 39). Although "living" and "organic," the molecules are not organized. The adjective "organic" is somewhat misleading, and the organic molecules do not possess any high degree of morphogenetic and organizational specificity. On the contrary, they need an "interior mold" to put them in the right places during ontogenesis, and to organize them into an animal or a plant.

This notion of "interior mold" is probably inspired by Bourguet. It relies on the idea that the body where the ontogenesis takes place (for example the body of a woman) has a certain material disposition that, through the action of "penetrating forces" equivalent to Newtonian attraction, organizes the organic molecules, inserts them in the right position by intussusception, and consequently shapes the new organism, just as a cake is shaped in a common mold, except that, here, the forces act on the whole volume and not only on the surface. The interior mold thus determines the morphogenesis, and its action continues throughout the life of the individual by assimilating the organic molecules brought by the food. In that respect, nutrition and reproduction are seen as the same process. As long as the organism grows, the organic molecules of the aliments are integrated to it, but, as soon as the growth is finished, they are available, as it were, to accumulate in genital organs and to form the seminal matter.

In the case of organisms capable of reproducing asexually (like plants or polyps), the interior molds correspond to all germs which make up the individuals and which all possess the information to build a complete individual (Buffon 1749–1767, 2: 55). But in organisms reproducing sexually, the interior mold is more complex and has to act on a mixture of organic molecules provided by both parents. Indeed, Buffon admits the existence of a female seminal matter, similar to the male sperm fluid, in order to make sense of the resemblance of each individual to both the mother and the father. These seminal liquids represent, according to him, "a sort of extract of all parts of the body" (Buffon 1749–1767, 2: 58). Once mixed, the molecules of the semens are immediately organized by the interior mold into a kind of new germ which grows afterwards, by intussusception of new molecules.

While the identification of the male semen does not pose any problem, the localization of the female semen is not so evident. Buffon guessed it could be in the cavity of the "globular body," that is, the Graafian follicle the ovists considered as the egg. To prove this hypothesis, he did a series of experiments with the collaboration of an eminent British scientist, John Turberville Needham, famous for his skill in microscopy (Sloan 1992). Their observations on the sperm fluid of man and several animals revealed the presence of a diversity of structures, instead of the animalcules Leeuwenhook had described some decades before. More surprisingly, they found the same kinds of structures in the so-called "female semen" of several animals (Buffon 1749–1767, 2: 203).

These results were astonishing, even in the context of eighteenth-century microscopy, all the more if they were compared to Leeuwenhook's observations. That is why many authors had doubts about their reality and suggested that Buffon and Needham had used bad microscopes, or were dazzled by their faith in their own theories. However, Sloan (1992) has shown that the quality of the microscopes was good. Maybe the material Buffon and Needham used was not fresh, and the structures they described were probably different kinds of microorganisms, or fragments of tissues from the dissected animals.

Anyway, Buffon interprets these results as proof of the similarity between the male and female semens. According to him, the various structures of the seminal fluids are not real "animalcules" but only transient clusters of organic molecules which are active enough to gather but, in the absence of any interior mold, are unable to form a true organism. He compares this phenomenon with the formation of microorganisms in culture fluids. In a sense, Buffon thus believes in spontaneous generation, but, for him, this process is fundamentally different from true generation, i.e., reproduction: "These moving bodies," he writes, "… are not produced by the ways of generation, they have no constant species, and, therefore, they can neither be animals nor plants" (Buffon 1749–1767, 2: 267).

On many occasions Buffon lays stress on this close link between the existence of living species and the faculty of reproducing eternally a constant form from one generation to the other through the action of the interior mold. He writes, for instance:

> It is from the union of these organic parts, returned from all parts of the body of the animal or the plant, that reproduction is accomplished, and it is always like the animal or plant in which it operates, because the union of these organic parts cannot be made but by the means of the interior mold, that is, in the order produced by the form of the body of the animal or the plant; in this consists the essence of the unity and continuity of the species; so these species will never been exhausted and will continue by themselves as long as their creator wants to let them survive.
>
> (Buffon 1749–1767, 2: 258)

The interior mold is thus the central concept, not only of Buffon's theory of generation but also of his whole conception of life and its diversity.

Buffon's alternative theory of generation in 1749

Up to this point, Buffon's theory of generation, as expounded in the first nine chapters of 1749, is consistent and supported by original (if strange) experimental data. However, the last two chapters of the *Histoire Générale des Animaux*, devoted to the formation, development, and growth of the fetus, present a set of theoretical reflections that, although not in conflict with the first chapters' ideas, seem to be independent from them.

Most strikingly, the notion of interior mold, which is so important in the main theory of 1749 and which would have been expected to play a major part in the description of embryogenesis and organogenesis, is nearly absent from these last chapters: here the term "moule" itself only appears in one sentence at the beginning of Chapter 10 ("On the formation of the fetus," Buffon 1749–1767, 2: 326), and then totally disappears.[7] This suggests that these texts may have been written before the rest of the *Histoire Générale des Animaux* and represent an earlier stage of Buffon's conceptions on generation, perhaps a vestige of his first system of the 1730s.

In these chapters, thus, Buffon attempts to make sense of the organization of the organic molecules, without using the interior mold. According to him, the organic molecules coming from one parent are incessantly moving; they can unite to form only transitory changing structures (such as the so-called "animalcules") but not true living beings. In order to give rise to a real ontogenesis, they need to be counterbalanced by different molecules, that is, by molecules coming from the genital parts of the other sex: dissimilar molecules represent, as it were, fulcra ("un point d'appui ou une espèce de base") for each other, so that their movement can stop and the molecules can gather and begin shaping an embryo (Buffon 1749–1767, 2: 336–342). The other organic molecules contained in the mixture of the semens and corresponding to the other parts of the body, that is, the molecules which are identical in the male and the female, join to this first "basis," "according to the laws of affinity which are between these different parts, and which determine the molecules to arrange themselves as they were in the individuals which furnished them" (Buffon 1749–1767, 2: 347). The position of each molecule in the embryo is similar to its previous position in the body of the parent (Buffon 1749–1767, 2: 329–330). Buffon is thus able to make sense of both the ontogenesis and the determination of the sex:

> [T]his mixture of organic molecules of the two individuals contains similar and different parts; the similar parts are the molecules which have been extracted from every part common to both sexes; the different parts are only those which have been extracted from parts distinguishing the male from the female. Thus there is, in this mixture, twice the number of organic molecules to form, for example, the head, or the heart, or any part common to both individuals, whereas there is only what is needed to form the parts of the sex. All similar parts (as the organic molecules of the parts common to both individuals are) can act on each other without being disturbed, and gather as

> if they had been extracted from the same body. But the dissimilar parts (as the organic molecules of the sexual parts are) cannot act on each other, nor mix closely, because they are not alike. Hence these parts alone will keep their nature without mixture, and will be the first to fix from themselves, without the need of being penetrated by the others; thus the organic molecules coming from the sexual parts will be the first fixed, and all the others, which are common to both individuals, will afterwards fix equally and indiscriminately, whether those of the male or those of the female, which will form an organized being perfectly resembling its father, if it is a male, or its mother, if it is a female, in the sexual parts, but which may resemble either the former, or the latter, or both, in all the other parts of the body.
>
> (1749–1767, 2: 330–331)

This theory explains several features of development. For example, the superfetation is the consequence of the overabundance of molecules able to form fulcra in the mixture of the semens: if two or many "fulcra" appear, and if the rest of the semens contains a sufficient quantity of organic molecules, two or many embryos can be formed. Furthermore, the sex of the fetus is determined by the proportion of male and female organic molecules in the mixture of semens: for instance, if there are more molecules coming from the male, the fulcrum is male. But, once the sex is determined, the rest of the body is predominantly formed by molecules coming from the other parent: as a consequence, boys resemble their mother more, and girls resemble their father (Buffon 1749–1767, 2: 342).

It would be conceivable, theoretically, that all this process of formation be totally or partially under the organizing control of an interior mold, but Buffon never mentions such an action in this part of the theory, and here the notion of interior mold itself seems to be completely unnecessary. In other terms, the morphogenetic information seems to rely entirely in the affinities of the organic molecules themselves, which is a considerable difference from the ideas presented in the first nine chapters, and an important common point with Maupertuis's first theory.

Buffon's views on the rhythm of the formation of the fetus are also different from those he presents in the two last chapters and in the rest of the *Histoire Générale des Animaux*. In the first nine chapters, the organization of the organic molecules by the interior mold is a nearly instantaneous phenomenon, and all parts of the organism are formed simultaneously. In a sense, this initial organization produces a kind of germ similar to the preexisting germs in the ovist theory, since it already possesses all the organs of the individual; the rest of the development is a mere growth of this first germ, by intussusception of other organic molecules (Buffon 1749–1767, 2: 43; on the comparison between Buffon's ideas on generation and the theories of preexistence, see Bowler 1973). Buffon unambiguously dismisses the notion of epigenesis and criticizes Harvey on this point:

> [W]e cannot doubt that the fetus is formed and exists as early as the first day, immediately after copulation, and consequently we must not give any credit to what Harvey says about the parts which come to fit each other by juxta-position, since, on the contrary, *they all exist from the outset*, and only unfold successively.
>
> (Buffon 1749–1767, 2: 292, emphasis added; see Cole 1930, 103; Roger 1989, 191)

This idea of the simultaneous and nearly instantaneous formation of parts is not absent from the last two chapters of the *Histoire Générale des Animaux*, and Buffon claims again that "everything is formed at the same time" (1749–1767, 2, 350). However, this does not concern all parts of the embryo any longer, and the rest of the development is no longer seen as a mere growth by intussusception of organic molecules. Now, Buffon recognizes three steps in ontogenesis: the initial formation we have presented below only produces an uncomplete sketch of the embryo, and it is followed by two stages of "development" (that is, in Buffon's sense, two kinds of complex unfolding-like processes). The first kind of development, which follows the initial formation of the embryo, is a complex process whose exact duration is not made clear, but it is not instantaneous. Unlike the last step, which is but a mere growth of already existing parts, it "is not a growth of all its composing parts at a constant ratio" but "a production of parts which seem to come into existence and which appear for the first time" (Buffon 1749–1767, 2, 366, 370).

Thus, even though the theory proposed in the last two chapters of the *Histoire Générale des Animaux* is not a true epigenesis in Harvey's sense, it is significantly more epigenetic than the theory as presented in the previous chapters, since, here, the initial formation of the embryo seems to be a short but noninstantaneous phenomenon, and the first kind of development, following the initial formation, corresponds to the *gradual* formation of new parts (see Roger 1989, 203–205).

Whereas the old question of the order of the formation of parts made no sense in the theory of the interior mold, Buffon deals with it in the last two chapters in an original way. Instead of merely looking for the first organ appearing in the embryo, he addresses the issue from both a mechanistic and an organismic standpoint. He admits that the first parts, formed by the physical laws of affinity, build the main axis of the body, corresponding to the vertebral column-to-be and, more generally, the single organs of the body (e.g., heart, gastrointestinal tract…), which are the most essential since the main functions of life depend on them. On the other hand, the double parts, being "more accessory and more external,"

> seem to derive their origin from the first ones, and to be made as much for the ornament, the symmetry and the external perfection of the animal, as for the necessity of its existence, and the exercise of the essential functions of life.
>
> (Buffon 1749–1767, 2, 367)

Thus, Buffon establishes a morphological and functional hierarchy of organs in which the central parts are more fundamental than the peripheral ones and have more significance in the general economy of the organisms as well as the classification of the animal groups. Half a century later, Cuvier will defend similar conceptions. But, at the same time, Buffon sketches a completely different, geometrical approach of animal morphology and morphogenesis. In his reflections on the symmetry of parts and its origin, he wonders how the double structures are initially folded together and how they gradually unfold. He compares these phenomena with the paper chains of children and wishes for the foundation of a new mathematical science to make sense of such processes:

> It is hardly possible to determine under what form the double parts exist before their development, how they are folded over each other, and what figure results from the position in relation to the single parts. The body of the animal, at the moment of its formation, certainly contains all the parts which are to compose it, but the relative position of these parts must be very different from what it becomes later. It is the same with every part of the animal or the plant, taken separately. If we observe the development of a budding leaflet, we shall see that it is folded on both sides of the main vein, that its lateral parts are as if they were superimposed, and that its figure does not resemble at all, at that time, the figure it is to take later. When we enjoy ourselves folding paper to produce afterwards, by a certain unfolding, regular and symmetric forms such as crowns, boxes, boats, etc., we can observe that the different folds we make in the paper seem to have nothing in common with the form which is to result from the unfolding. We only see that the folds are always made in a symmetrical order, and that we make on one side what we have just made on the other. But it would be a problem beyond known geometry to determine the figures which can result from every unfolding of a certain number of given folds. Our mathematical sciences totally lack everything what bears some immediate relation to the position; this art, which Leibniz called *Analysis situs*, is not yet born. And yet, this art, which would make us know the relations of position between things, would be as useful to the natural sciences as the art whose object is only the dimensions of things, and it would be perhaps more necessary than it, because we need more often to know the form than the matter.
>
> (Buffon 1749–1767, 2: 372–374)

Here the distinction between symmetric and asymmetric parts recalls one of the few elements Buffon gives on the first theory of generation he had imagined in the 1730s (see below), which suggests that these two last chapters of 1740 are a remnant of an earlier stage of his thought, and that he has integrated these texts into his main theory of 1749 without introducing the notion of interior mold into them. Indeed, even though the action of an interior mold is possible (but not required) during the initial formation of the embryo, it is barely conceivable during the two steps of "development."

Consequences of Buffon's theory of generation of 1749

The theory proposed in 1749 in the *Histoire Générale des Animaux* is thus composed of at least two elements, which are not contradictory but are not very well connected to each other. From a conceptual standpoint, both are original and virtually fruitful.

The second one, which is presented in the two last chapters and may be a vestige of Buffon's earlier conceptions, is less known than the first, because it was not developed further in the *Histoire Naturelle*. It has a Newtonian flavor and shares some features with Maupertuis's first system of 1745 since, as we have seen, the organization of the organic molecules is determined by forces of affinity similar to the gravitation, and no higher entity such as an interior mold seems to act on them. Buffon's theory is certainly much more complex than Maupertuis's, but in both cases, the explanation of ontogenesis makes possible all kinds of "mistakes" in the reproduction, and could have led Buffon, like Maupertuis, to daring ideas on the variations of species.

However, in all other chapters of the *Histoire Générale des Animaux*, Buffon defends the theory of the interior mold, which is also favored in the following parts of his work (especially in the volumes devoted to quadrupeds), and all of his conception of species and its alterations is determined by this choice.[8] In this theory, some plasticity certainly remains possible: climate, food, and other external factors can interfere during the processes of generation and sometimes bring significant changes within a certain species, which makes sense, for instance, of the existence of varieties in the human species as well as in domestic and wild animals. In that respect, Buffon's theory allows for more alteration of species than the conception of preexisting germs. But the organizing action of the interior mold is fundamentally reliable, and it ensures the constancy of the specific form from one generation to the other. As a consequence, there is no transformation of one species into another, and it is impossible to obtain fertile hybrids between two distinct natural species. Even though Buffon often hesitated, over the years, as to the exact range of certain species (for example, whether the donkey and the horse belong to the same species), he never changed his general conception of species, whose keystone was the notion of interior mold. Just as Maupertuis's purely epigenetic system led to transformism, Buffon's dismissal of true epigenesis prevented him from accepting the true evolution of species.

In the theory of the last two chapters of the *Histoire Générale des Animaux*, as well as in Maupertuis's views of 1745, the main information for morphogenesis lies in the forces between the organic molecules, whereas, in the theory presented in the first chapters, this information dwells in the interior mold. If we admit that the last chapters correspond to earlier views, is it interesting to compare the different evolution of Buffon's and Maupertuis's ideas on generation. As we have seen, after 1745, Maupertuis acknowledges a higher degree of autonomy and specificity of the particles of the semen, and he even imagines that they could have a kind of "instinct," which could explain their organization. In Leibnizian terms, he attributes to the particles some features of monads.

Buffon totally dismisses this idea of an "instinct" of organic molecules: "We shall not say, with some philosophers, that matter, under whatever form it may be, knows its existence and relative faculties" (1749–1767, 2: 3). Although he thinks, like Maupertuis, that the organic molecules coming from one organ of the father or the mother participate in the formation of the same organ in the embryo, he does not emphasize this point, and he considers that the organic molecules are essentially nonspecific: indeed, they follow an eternal cycle of unions and dispersions, and are able to pass from one body to another, not necessarily of the same species. Their organization into a living being does not result from their intrinsic nature and properties but from the action of a factor external to them, namely the interior mold, which possesses the morphogenetic information and ensures its transmission throughout the generations. In the dialectic between the organizing mold and the organized molecules, the former clearly has the primacy. Buffon writes in 1756: "What is most constant, most unchanging in Nature is the imprint [*l'empreinte*] or the mold of each species, both in animals and in plants; what is most variable and perishable [*corruptible*] is the substance of which they are composed" (1749–1767, 6: 86–87).

Maupertuis gradually shifted from a Newtonian model, where the forces between the elementary particles composing the organism particles played the major part in ontogenesis, to a kind of Leibnizianism, where the monads were virtually identified with these elementary particles. Buffon too started from a Newtonian theory and moved to a sort of Leibnizianism, but in a very different way. On the one hand, he retained some elements of Newtonianism, since he conceived of the action of the interior mold as a set of penetrating forces. On the other hand, unlike Maupertuis, he rejected the idea of an autonomous morphogenetic faculty of the organic molecules, which had nothing to do with monads, although they were "living." He preferred to ascribe this organizing faculty to the interior mold, to which he gave, in a sense, some features of the monad, even though, paradoxically, its action was rather Newtonian. This conception was closer to Leibniz's own interpretation of the monad (Smith 2011), since the interior mold corresponded to the whole organism rather than to its elementary components.

Beyond Leibnizianism, Buffon's reflections on the action of the interior mold on organic molecules have some Aristotelian character. For example, as he explains that the organic molecules brought by nourishment can retain some properties of their previous vegetal state instead of being completely assimilated in their new organism (e.g., in the antlers of a deer), he uses the Aristotelian notions of form and matter:

> Matter, in general, seems to receive every form indiscriminately, and to be able to bear every possible imprint: the organic molecules, that is, the living parts of this matter, pass from plants to animals without being destructed nor altered, and equally form the living substance of grass, wood, flesh or bones. Thus it seems, at this first glance, that matter can never predominate over form.... However, by observing Nature in a more specific way, we

> shall realize that these organic molecules are sometimes not perfectly assimilated to the interior mold, and that matter has a quite noticeable influence on form.... Therefore, this organic matter the animal assimilates to its body by the nutrition is not completely indifferent to any modification, it is not completely deprived from its previous form, and it retains some characters of the imprint of its first state.
>
> (1749–1767, 6: 87–88)

Strikingly, the dialectics between matter and form is very similar to Aristotle's own views on this issue, with the same subtleties on the possible resistance of matter to form.

The evolution of Buffon's ideas on generation and organic molecules after 1765

Whether a Leibnizian monad or an Aristotelian form, the interior mold was the central concept in the main theory Buffon proposed in 1749. Since all the morphogenetic information belonged to molds, they predominated over organic molecules and represented a key link between the problem of generation and the question of species and its constancy. This connection remained in later texts, but the theory significantly evolved after the mid-1760s. In particular, the question of the origin of the interior molds and the possible formation of new molds in the history of the Earth emerged in Buffon's writings, which led to an inversion of the hierarchy between the mold and the organic molecules.

In 1765, in the second *Vue de la Nature*, Buffon suggests that the organic matter would be virtually able to be incorporated into new molds if its quantity was sufficient:

> If [organic matter] were overabundant, if it were not, at all times, equally employed and entirely absorbed by the existing molds, other molds would be formed, and we would see new species appear, because this living matter cannot remain idle, because it is always active, and it has only to combine with brute parts to form organized bodies.
>
> (1749–1767, 13: ix)

Twelve years later, Buffon goes further, and he acknowledges that the organic molecules have the faculty to "knead" (*travailler*) raw material and even, under certain circumstances, to form new interior molds spontaneously:

> Let us suppose, for a while, that the Sovereign Being would like to remove life from all existing individuals, that all of them were struck by death at the same time. The organic molecules would certainly outlive this universal death. Since the number of these molecules would always be the same, and since their indestructible essence would be as permanent as the essence of the brute matter which nothing would have destroyed, Nature would always

> possess the same quantity of life, and we would see new species appear which would replace the previous ones; for the living organic molecules, being all free and neither soaked up nor absorbed by any existing mold, could work brute matter on a large scale, and produce, at first, an infinity of organized beings, among which some would only have the faculty of growing and feeding, while others, more perfect, would be endowed with the faculty of reproducing.... The power of these organic molecules being proportionate to their number and freedom, *they would form new interior molds*, to which they would give all the more extent as they were in larger quantities to participate in the formation of these molds, which, consequently, would represent a new living Nature, perhaps quite similar to the Nature we know.
>
> (Buffon 1774–1789, 4: 359–363, emphasis added; see also 5: 184–186)

Here the organic molecules have an organizational power they did not have in 1749, since they can produce molds spontaneously. In the final analysis, the whole organization of the living world, including the constant form of living species, results from the properties of the organic molecules. Technically, there is no overt contradiction between these ideas of 1777 and the *Histoire Générale des Animaux*, since in 1749 nothing was said about the initial formation of the interior mold. But, by giving an explanation of the origin of the molds, and by adding a historical dimension to his theory of reproduction, Buffon completely changes the status of the molds and the molecules, in favor of the latter.

At first sight, this evolution of Buffon's thought could be seen as resulting from the influence of vitalist trends, or from a kind of Leibnizianism similar to Maupertuis's second theory (see above): in this interpretation, he would have transferred the monadic character from the interior mold to the molecules. But this is not the case, and, in 1777 as in 1749, Buffon, dismisses the notion of an "instinct" or a "memory" of the organic molecules. As a matter of fact, the exact nature of their morphogenetic activity is not clear, but it seems to be unspecific since Buffon considers them the result of a very general physical phenomenon, namely the action of the active element of nature (i.e., heat, light, or fire, which are the same thing) on the passive elements (water, earth, and air). According to him, each atom of light can "shake and penetrate" many atoms of passive elements; the consequence of this impulsion given by the heat, combined with the attractive force "common to all parts of matter," is that "each brute and passive atom becomes active and living when it is penetrated in all its dimensions by the vitalizing element." Heat is thus "the primitive element of life" (Buffon 1774–1789, 4: 365–366; see also 5: 115).

In the *Histoire Naturelle des Minéraux*, Buffon further explains that, by the combined action of heat and attraction, the "ductile matter ... takes the form of an organized germ which will soon become living or vegetating [*un germe organisé, qui bientôt deviendra vivant ou végétant*], through its continuous development and its proportional extension in length, width and depth" (Buffon 1783–1788, 1: 5). The only difference between living beings and minerals is the more or less "ductile" character of the brute matter on which heat and attraction act: if this matter is not ductile enough, it cannot be penetrated in its whole volume and become living, and it can only produce mineral molecules. If it is soft enough, it can become a living "germ" or interior mold, which, by the addition of other organic molecules, grows and reproduces.

There is no antinomy between these views and those presented in earlier texts, but the introduction of historical considerations leads Buffon to be more specific as to the physical causes at the origin of life. In the later version of his theory of generation, the interior mold is a mere step in a process fundamentally driven by the activity of organic molecules, which themselves result from the action of very general and unspecific forces on the brute matter. This is an important shift since, in 1749, the interior molds, of which the origin and formation were not considered at all, could be (at least virtually) conceived in an essentialist way and compared to a monad or even an Aristotelian form. In the 1770s, there was nothing in common between the Leibnizian monads and the interior molds, contingently produced by Newtonian-like forces.

This difference between the two states of the theory, in 1749–1760 and after 1765, may be explained by nonscientific factors, since it would be credible that, in 1749, Buffon cautiously avoided addressing the issue of the historical origin of the interior molds in a materialistic way, but felt free to do it two decades later, in a less repressive context. But another possible interpretation is that his approach really changed in the meantime: this is all the more plausible considering that Buffon's epistemology of 1749 established a gap between scientific *theories*, restricted to the study of constant causes and repeated events (e.g., the nonhistorical theory of the Earth presented in the *Second Discours*: Buffon 1749–1767, 2: 63–124), and merely hypothetical *systems* on unique historical events such as the formation of the planets (ibid., 127–167). This distinction had disappeared in the 1770s, by which time texts such as the *Époques de la Nature* were published, and the epistemic status of the historical approach was no longer lower (on this point, see Roger 1989: 528–558).

Leibnizianism certainly played an important part in Buffon's thought from the 1740s (Sloan 1979), but, with regard to the theory of generation, it may have been a mere step in the evolution of his ideas from a first purely Newtonian system, elaborated in the 1730s, to a revised Newtonian conception enriched with the historical dimension in the 1770s. Whatever cause led Buffon to be more specific in the 1770s as to the formation of the organic molecules and the interior molds, its philosophical consequences were considerable since it made possible to conceive a strictly materialistic explanation, not only of the generation of the living beings but of life itself and its origin on Earth.

Notes

1 On the problem of generation in the early modern period, see e.g., Needham 1959, Roger 1993 [1963]; Adelmann 1966; Bowler 1971; Roe 1981; Wilson 1997; Pinto-Correia 1998; Smith 2006.

2 Indeed, Harvey himself (1651) coined the term "epigenesis." On his theory, see French 1994; Keller 1999; Lennox 2006; Ekholm 2008.
3 On Bourguet, see Schiller 1975; Ibrahim 1987; Duchesneau 2003; Cheung 2006; Schmitt 2014.
4 On Maupertuis's theory of generation, see Brunet 1929, Chapter 7; Glass 1959; Roger 1993: 468–487; Duchesneau 1982: 236–258; Hoffheimer 1982; Sandler 1983; Beeson 1992; Terrall 2002: 202–230; 2007; Wolfe 2010; Schmitt 2014.
5 On Buffon's theory of generation, see e.g., Roger 1993: 542–558; 1989: 166–207; Castellani 1972; Bowler 1973; Dougherty 1980, Chapter 6; Duchesneau 1982: 268–277; 2009; Kubbinga 1990; Reill 2005: 33–70; and the references mentioned below.
6 On Buffon's concept of species, see Farber 1972; Bowler 1973; Sloan 1976; 1979; 1987; Gayon 1992.
7 It reappears only in the brief "Recapitulation" at the end of the *Histoire Générale des Animaux* (420–426). By contrast, the term "organic molecule" is frequent in Chapter 10.
8 On the question of transformism in Buffon's thought, see e.g., Wilkie 1956; Lovejoy 1959; Farber 1972; Sloan 1976; 1979; 1987; Barsanti 1984; 1992; Eddy 1984; Roger 1989: 378–441; Gayon 1992.

References

Adelmann, H. B. 1966. *Marcello Malpighi and the Evolution of Embryology*. 5 vols. Ithaca, NY: Cornell University Press.

Aucante, V. 2006. Descartes's experimental method and the generation of animals. In J. E. H. Smith (ed.), *The Problem of Animal Generation in Early Modern Philosophy* (pp. 65–79). Cambridge: Cambridge University Press.

Barsanti, G. 1984. Linné et Buffon: deux visions différentes de la nature et de l'histoire naturelle. *Revue de Synthèse, 3rd series, 113–114*, 83–111.

Barsanti, G. 1992. Buffon et l'image de la nature: de l'échelle des êtres à la carte géographique et à l'arbre généalogique. In J. Gayon (ed.), *Buffon 88, Actes du Colloque international pour le bicentenaire de la mort de Buffon* (pp. 255–296). Paris: Vrin.

Beeson, D. 1992. *Maupertuis: An Intellectual Biography*. Oxford: Voltaire Foundation.

Bourguet, L. 1729. *Lettres philosophiques sur la formation des sels et des crystaux, et sur la Génération et le Mechanisme Organique des plantes et des animaux*. Amsterdam: L'Honoré.

Bowler, P. J. 1971. Preformation and preexistence in the seventeenth century: A brief analysis. *Journal of the History of Biology*, *4*, 221–244.

Bowler, P. J. 1973. Bonnet and Buffon: Theories of generation and the problem of species. *Journal of the History of Biology*, *6*, 259–281.

Brunet, P. 1929. *Maupertuis. L'œuvre et sa place dans la pensée scientifique et philosophique du XVIIIe siècle*. Paris: Blanchard.

Buffon, G. L. Leclerc de. 1749–1767. *Histoire naturelle, générale et particulière, avec la description du Cabinet du Roi*. 15 vols. Paris: Imprimerie Royale.

Buffon, G. L. Leclerc de. 1774–1789. *Histoire naturelle. Supplément*. 7 vols. Paris: Imprimerie Royale.

Buffon, G. L. Leclerc de. 1783–1788. *Histoire naturelle des Minéraux*. 5 vols. Paris: Imprimerie Royale.

Buffon, G. L. Leclerc de. 1971. *Correspondance générale*. Ed. H. Nadault de Buffon. 2 vols. Geneva: Slatkine Reprints.

Buffon, G. L. Leclerc de 2007. *Œuvres complètes*. Eds. S. Schmitt and C. Crémière. 10 vols to date. Paris: Champion.

Castellani, C. 1972. The problem of generation in Bonnet and in Buffon: A critical comparison. In A. G. Debus (ed.), *Science, Medicine and Society in the Renaissance. Essays to Honor Walter Pagel* (pp. 265–288). New York: Science History Publications.

Cheung, T. 2006. The hidden order of preformation: Plans, functions, and hierarchies in the organic systems of Louis Bourguet, Charles Bonnet and Georges Cuvier. *Early Science and Medicine, 11*, 11–49.

Cole, F. J. 1930. *Early Theories of Sexual Generation*. Oxford: Clarendon.

Dougherty, F. W. P. 1980. *La métaphysique des sciences. Les origines de la pensée scientifique et philosophique de Buffon en 1749*. PhD dissertation, Université de Paris I.

Duchesneau, F. 1982. *La Physiologie des Lumières. Empirisme, Modèles et Théories*. The Hague: Nijhoff.

Duchesneau, F. 2003. Louis Bourguet et le modèle des corps organiques. In M. T. Monti (ed.), *Antonio Vallisneri. L'edizione del testo scientifico d'Étà moderna* (pp. 3–31). Florence: Olschki.

Duchesneau, F. 2009. Buffon et l'épigenèse. In M.-O. Bernez (ed.), *L'Héritage de Buffon* (pp. 273–293). Dijon: Éditions Universitaires de Dijon.

Eddy, J. H. 1984. Buffon, organic alterations, and man. *Studies in History of Biology, 7*, 1–45.

Ekholm, K. J. 2008. Harvey's and Highmore's accounts of chick generation. *Early Science and Medicine, 13*, 568–614.

Farber, P. L. 1972. Buffon and the concept of species. *Journal of the History of Biology, 5*, 259–284.

French, R. 1994. *William Harvey's Natural Philosophy*. Cambridge: Cambridge University Press.

Gayon, J. 1992. L'individualité de l'espèce, une thèse transformiste? In J. Gayon (ed.), *Buffon 88, Actes du Colloque international pour le bicentenaire de la mort de Buffon* (pp. 475–489). Paris: Vrin.

Glass, B. 1959. Maupertuis, pioneer of genetics and evolution. In B. Glass, O. Temkin, and W. L. Strauss (eds.), *Forerunners of Darwin: 1745–1859* (pp. 51–83). Baltimore, MD: Johns Hopkins Press.

Hanks, L. 1966. *Buffon avant l'"Histoire Naturelle."* Paris: Presses Universitaires de France.

Harvey, W. 1651. *Exercitationes de generatione animalium. Quibus accedunt quædam De Partu: de Membranis ac humoribus Uteris: & de Conceptione*, London: Pulleyn.

Hoffheimer, M. H. 1982. Maupertuis and the eighteenth-century critique of preexistence. *Journal of the History of Biology, 15*, 119–144.

Ibrahim, A. 1987. La notion de moule intérieur dans les théories de la génération au XVIIIe siècle. *Archives de Philosophie, 50*, 555–580.

Keller, Eve. 1999. Making up for losses: The working of gender in William Harvey's *De Generatione Animalium*. In S. Greenfield and C. Barash (eds.), *Inventing Maternity: Politics, Science, and Literature 1650–1865* (pp. 34–57). Lexington, KY: University of Kentucky Press.

Kubbinga, H. H. 1990. Les origines de la théorie cellulaire; les "molécules organiques" de Buffon. *Centaurus. International Magazine of the History of Mathematics, Science, and Technology, 33*, 175–213.

Lennox, J. G. 2006. The comparative study of animal development. William Harvey's Aristotelianism. In J. E. H. Smith (ed.), *The Problem of Animal Generation in Early Modern Philosophy* (pp. 21–46). Cambridge: Cambridge University Press.

Lovejoy, A. O. 1959. Buffon and the problem of species. In B. Glass, O. Temkin, and W. L. Strauss (eds.), *Forerunners of Darwin: 1745–1859* (pp. 84–113). Baltimore, MD: Johns Hopkins Press.

Maupertuis, P. L. Moreau de. 1745. *Vénus physique*. S. l.: s. n.

Maupertuis, P. L. Moreau de. 1768 [1751]. Système de la nature. In *Œuvres de Maupertuis. Nouvelle édition* (4 vols.), vol. 2, pp. 139–184. Lyon: Bruyset.

Needham, J. 1959. *A History of Embryology*, 2nd ed. Cambridge: Cambridge University Press.

Pinto-Correia, C. 1998. *The Ovary of Eve: Egg and Sperm and Preformation*. Chicago, IL: University of Chicago Press.

Reill, P. H. 2005. *Vitalizing Nature in the Enlightenment*. Berkeley, CA: University of California Press, 2005.

Rey, R. 1992. Buffon et le vitalisme. In J. Gayon (ed.), *Buffon 88, Actes du Colloque international pour le bicentenaire de la mort de Buffon* (pp. 399–413). Paris: Vrin.

Rey, R. 2000. *Naissance et développement du vitalisme en France de la deuxième moitié du 18e siècle à la fin du Premier Empire*. Oxford: Voltaire Foundation.

Roe, S. A. 1981. *Matter, Life, and Generation. Eighteenth-Century Embryology and the Haller-Wolff Debate*. Cambridge: Cambridge University Press.

Roger, J. 1989. *Buffon. Un philosophe au Jardin du Roi*. Paris: Fayard.

Roger, J. 1993 [1963]. *Les sciences de la vie dans la pensée française du 18e siècle*. Paris: Albin Michel.

Sandler, I. 1983. Pierre Louis Moreau de Maupertuis – A precursor of Mendel? *Journal of the History of Biology*, *16*, 101–136.

Schiller, J. 1975. La notion d'organisation dans l'œuvre de Louis Bourguet (1678–1742). *Gesnerus*, *32*, 87–97.

Schmitt, S. 2014. Mécanisme et épigenèse: les conceptions de Bourguet et de Maupertuis sur la génération. *Dix-huitième siècle*, *46*, 131–154.

Sloan, P. R. 1976. The Buffon-Linnaeus controversy. *Isis*, *67*, 356–375.

Sloan, P. R. 1979. Buffon, German biology, and the historical interpretation of biological species. *British Journal for the History of Science*, *12*, 109–153.

Sloan, P. R. 1987. From logical universals to historical individuals: Buffon's idea of biological species. In *Histoire du concept d'espèce dans les sciences de la vie (Colloque international, Paris 1985)*, pp. 101–140. Paris: Fondation Singer-Polignac.

Sloan, P. R. 1992. Organic molecules revisited. In J. Gayon (ed.), *Buffon 88, Actes du Colloque international pour le bicentenaire de la mort de Buffon* (pp. 415–438). Paris: Vrin.

Smith, J. E. H. (ed.). 2006. *The Problem of Animal Generation in Early Modern Philosophy*. Cambridge: Cambridge University Press.

Smith, J. E. H. 2011. *Divine Machines: Leibniz and the Sciences of Life*. Princeton, NJ, and Oxford: Princeton University Press.

Terrall, M. 2002. *The Man Who Flattened the Earth. Maupertuis and the Sciences in the Enlightenment*. Chicago, IL: University of Chicago Press.

Terrall, M. 2007. Speculation and experiment in Enlightenment life sciences. In S. Müller-Wille and H. J. Rheinberger (eds.), *Heredity Produced. At the Crossroad of Biology, Politics, and Culture, 1500–1870* (pp. 253–275). Cambridge, MA: MIT Press.

Wilkie, J. S. 1956. The idea of evolution in the writings of Buffon. *Annals of Science*, *12*, 48–62, 212–227 and 255–266.

Wilson, C. 1997. *The Invisible World. Early Modern Philosophy and the Invention of the Microscope*. Princeton, NJ: Princeton University Press.

Wolfe, C. T. 2010. Endowed molecules and emergent organization: The Maupertuis-Diderot debate. *Early Science and Medicine*, *15*, 38–65.

Wolfe, C. T., and Terada, M. 2008. The animal economy as object and program in Montpellier vitalism. *Science in Context*, *21*, 537–579.

3 Metaphysics and "vital" materialism

Émilie Du Châtelet and the origins of French vitalism

Phillip Sloan

Introduction

Developments under consideration in this chapter in part set up the framework for what was to emerge around 1800 as the domain of a separate science – *Biologie* – devoted to a "living" world distinct from that studied by the physical sciences. This development, magisterially analyzed in John Zammito's recent study (Zammito 2018), rested upon several new claims – claims about special "laws of life," appeals to new forces not found in the physical domain, new epistemic claims about the validity of appeals to efficient causes inferred abductively from phenomena, and distinctions of "living" from "dead" matter.

In a longer historical perspective, there is nothing deeply novel in itself about a distinction of the living from the nonliving. It is integral to the Aristotelian tradition and the classical heritage of Greek medicine. But these alternatives had been "eclipsed" – I use this metaphor intentionally – by the rise of strong biomechanism in the latter seventeenth century, particularly developing upon René Descartes's foray into the life sciences in his published and posthumous writings, assisted by contributions from disciples of Galileo and some interpretations of Newton.

The "re-emergence" of a science of "life" in the middle decades of the eighteenth century, typically in reaction against the strong mechanism of the Dutch, Italian, and Scots iatromechanists who had dominated much of "official" medical theory and teaching in the late seventeenth century, complexifies the common narrative of the Scientific Revolution in regard to the life sciences (Wolfe 2011; Distelzweig *et al.* 2016; Zammito 2018). These developments assisted in opening up new horizons that grounded what Peter Hans Reill has termed the "vitalist" transformation of Enlightenment thought (Reill 2005). With these foundational changes, the conditions were set for a new "science" of the living state itself.

In this chapter it is my intent to deal with a select set of issues in these transformations leading to the development of "vital materialism" as this emerges to prominence in the 1740s, particularly in French life science circles. My approach to these issues is novel and has not been discussed by recent reviews of these developments by Zammito (2018), Gaukroger (2010), and others who have examined aspects of this profound development. By this designation, I refer to two alternative meanings of "vital" materialism. One version was articulated by those who held to the metaphysical claim that special causal powers were inherent in matter itself and these causes produced the living phenomena of embryonic development, tissue repair, regeneration, and the other functional activities of organisms. It is important to see that this was unlike classic Aristotelian hylomorphism, since it lacked the teleological dimensions of the Aristotelian view of the life–matter relation. It is best classed as a form of hylozoism. This was justified by some on the authority of Newton and his introduction of what many took to be inherent microforces within matter in the Thirty-First Query to the *Opticks*. Those endorsing at least aspects of this approach might also be grouped within the "experimental" Newtonian tradition recently analyzed by Zammito (2018, Chapter 2).

This version can be distinguished from the views developed by Montpellier medical theorists who were not technically "vital materialists" in this sense, since they held views more correctly characterized as a "prudent positivism" (Wolfe 2014, 238) that rescinded from a metaphysical level of explanation, positing efficient causes inferred abductively from the phenomena. This was also at times justified in the name of the great Newton.

The strand I will detail can be distinguished from both of these, although it leads more directly in the direction of the first meaning above, i.e., toward hylozoism. I will detail how an originally metaphysically grounded teleological theory of life, resting upon a conception of dynamic force as more fundamental than matter, was transformed into a thesis about the constitution of living beings through their construction from a special kind of vital matter possessed of special intrinsic powers. This transition takes place in different phases, and this chapter will be restricted to its impact on the intermediate position represented by the theory of the *molécules organiques* of George Louis Comte de Buffon (1709–1788). Extension of this theme would also lead us into the views of Julien Offray de La Mettrie (1709–1751) who was also in touch with these same developments. But La Mettrie took these ingredients further in the direction of ontological vital materialism. I shall approach this from a previously unexplored direction, specifically through the examination of the impact of the work of the French *philosophe* Gabrielle Émilie Le Tonnelier de Breteuil, Marquise Du Châtelet (1706–1749) on Buffon's reflections on the relations between life, matter, and organic function.

Émilie Du Châtelet is best known for her important synthesis and reinterpretation for a French audience of the philosophical concepts of Gottfried Leibniz as they were refracted through the teachings of Christian Wolff, bringing Leibnizianism into a creative dialogue with Newtonian mechanics (Detlefsen 2014; Shank 2008, Chapter 7; Janik 1982; Iltis 1977; Barber 1967, 1955). Her synthesis of these issues, first published anonymously in 1740 in Paris as the *Institutions de Physique*, with a second edition issued in Amsterdam under her own name in 1742 that became the basis of translations into German and Italian in 1743, served as a primary vehicle by which a synthetic interpretation of the

Leibnizian–Wolffian philosophy reached a wider French intellectual audience. Her impact was also indirect through the articles by Samuel Formey in the *Encyclopédie*, which often reflect extensive borrowing directly from her work without acknowledgment.[1]

The primary philosophical dimensions of her synthesis were laid out in the first nine chapters of the *Institutions*. These are based on her own readings of works of Christian Wolff and insights she obtained from the private tutorials conducted at her residence in Cirey in the spring of 1739 by the Hessian philosopher and disciple of Wolff's philosophy Johann Samuel Koenig (1712–1757). Aspects of the Wolffian philosophical program also enter into her final revisions of the additional 12 chapters of the *Institutions* on mechanics, creating an unusual work that explored the deeper metaphysical foundations of physics and philosophy. Although questions of mechanics predominate in her work, I will seek to show how her work is important in subtle and indirect ways for some of the major transformations in biomedical theory that emerged in the 1740s.

My exposition will proceed in three parts. First, I will discuss some of the background issues in the Leibniz–Wolff philosophy concerning the relation of living phenomena to a metaphysics of force. This will be connected to a theory of matter that was then extended to deal with living organisms. Second, I will analyze Du Châtelet's restatements of these principles in her *Institutions*. Third, I will illuminate how the direct contact with Du Châtelet's views can be seen to impact upon the views of life and matter developed in the 1740s by Buffon and his theory of the *molécules organiques*.

Leibnizian organic machines

The early 1740s represent a period when certain key Enlightenment *philosophes* were searching for a way to complete closure on a naturalistic view of the world. But it is clear that, until the 1740s, the problem with completing this program was always the question of organic beings. If organisms were simply divinely designed material machines, as claimed by advocates of the mechanical philosophy, the origin of such machines seemed resolvable only by assumptions of a strong theocentric creationism that underpinned the generally accepted preexistence embryology. Whether interpreted in its ovist, vermist, or "preexistent germ" formulations,[2] the question of organic origins and the thesis of biomechanism were deeply linked.

The main outlines of the generation debates of the early eighteenth century have been described by a substantial literature and can be presumed (Smith 2006; Pinto-Correia 1997; Roger 1994; Roe 1981). As the preferred theory of generation promulgated by biomechanists, preexistence theory, in one of its versions, formed the paradigmatic account of embryological formation expounded by biomedical writers of the period and taught in the major texts and the most common *Institutes of Medicine*.

The interaction of this theory with the philosophical program of Gottfried Leibniz and his interpreters, especially those made by the University of Halle professor Christian Wolff, has also been the subject of a substantial literature, and only the highlights will be addressed here (Phemister 2011, 2005; Smith 2011; Duchesneau 2011, 2010). Leibniz factored the problem of generation into his larger philosophical structure – the principle of the preestablished harmony, the metaphysical theory of dynamic monads as more ultimate than matter, and the directional teleology supplied by the principle of sufficient reason. These complex developments, through the intermediary of his synthesizer and disciple Christian Wolff, form the thread we will follow into Du Châtelet's work.

As this was relevant to living beings, Leibniz developed on his principles a complex theory of "natural machines" that on one hand embraced preformationist embryology but on the other moved beyond the blunt mechanism that had accompanied this theory in the biomedical literature of the period. In his 1695 *Specimen Dynamicum* and other writings in this period, Leibniz introduced a distinction of levels of forces: the "primitive force" (*vis primitiva*) that resides in all "corporeal" substance as its primary entelechy or even substantial form, and the "derivative force" (*vis derivata*) that is manifest in secondary matter (Phemister 2005, Chapter 8).

In the reflections built upon this, he drew explicit connections between his dynamic theory of substance, the concept of the preestablished harmony, and the reigning preexistence embryology embraced by biomechanists, using this connection to deal with both the problem of ensoulment and also the source of biological order. This also drew upon his awareness of the empirical work and theoretical assumptions of microscopists of the late seventeenth and early eighteenth centuries, especially Marcello Malpighi, Jan Swammerdam, Nicholaas Hartsoeker, and Anton van Leeuwenhoek.

Leibniz was unequivocal in his endorsement of the strong theory of an encasement of forms reaching back to the first origins of the world in divine creation. The preformed beings simply unfold mechanistically and the organism for all practical purposes is a causal mechanism. As he writes in the fifth letter of the Leibniz–Clarke correspondence:

> The organism of animals is a mechanism, which supposes a divine preformation. What follows upon it, is purely natural, and entirely mechanical. Whatever is performed in the body of man, and of every animal, is no less mechanical, than what is performed in a watch. The difference is only such, as ought to be between a machine of divine invention, and the workmanship of such a limited artist as man may be.
>
> (Leibniz in Alexander 1984, 93)

This Leibnizian endorsement of a strong theory of literal *emboîtement*, with organisms infinitely encased within one another, was not the only way this Leibnizian approach was interpreted by his successors. Of relevance to our discussion is a weaker theory of generation that can be encountered in authors generally impacted by Leibnizianism, and is the one exploited by Du Châtelet. This is the more flexible theory of the preexistent *germes* in which only primordial germs, capable of becoming full organisms under the proper preordained conditions, are assumed to be preformed.[3] This hypothesis, introduced into the discussions by the French architect and naturalist Claude Perrault in his *Méchanique des Animaux* in 1680 and republished in his *Œuvres Diverses* in 1721, escaped some of the difficulties of literal *emboîtement*. As Perrault proposed this hypothesis, the original creation was comprised of two kinds of bodies, one simple and similar in its components – the fundamental "elements" – and the other "composite and organic." The latter are "bodies capable of life, furnished with all the organs necessary for their function, but so minute that it is impossible for them to exercise these."

> [I]n this state being mixed among inanimate bodies, they await the occasion of encountering a substance sufficiently subtle and arranged such that it can penetrate the narrow conduits of their little organs, and make them proper to receive nourishment which will make them acquire a proper magnitude. Then they have reached what is called generation.
>
> (Perrault 1721, II, Chapter 9, 482. All translations my own unless otherwise noted)

In the Leibnizian reading of relevance to our discussion, the organic "machine" comes to be identified with the preformed *germes*. These supervene on a nonmaterial and dynamic substance. Such machines are forms of corporeal *matter*, which are, however, underlain ultimately by the immaterial monads which constitute the deeper *substance* of the organism. As Leibniz put this in a draft letter to Thomas Burnet in 1699 in which he engaged the controversy over whether or not attractive force is essential to matter, he argued:

> In bodies, I distinguish the corporeal substance from the matter, and I distinguish first from secondary matter. The secondary matter is an aggregate or composite of several corporeal substances, as a flock [*trouppeau*] is a composite of several animals. But each animal and each plant is also a composite substance, having in itself the principle of unity, which makes it truly a substance and not an aggregate.
>
> (Leibniz 1699 in Gerhardt 1875, III, 260)

As we see, this composite body cannot be considered to be only an interconnected and interacting aggregate of dynamic parts, as might be conveyed by the popular metaphor of the "bee swarm" employed by vitalist physicians and popularized by Diderot in the *Rêve de D'Alembert* (Diderot 1956, 111). Each organized being comprised of this composite of secondary matter has organizing and directing it a dominant immaterial and teleologically oriented monad that unifies it. As Leibniz continues in this letter,

> this principle of unity is what one calls soul, or properly something with analogy to the soul. But beyond the principle of the unity, corporeal

substance has its mass or secondary matter, which is still an aggregate of other smaller corporeal substances, and this goes on to infinity.

(Ibid.)[4]

This Leibnizian distinction of "substance" from "matter," combined with the underlying metaphysics of force, give us the needed ingredients of the Leibnizian theory we will explore below. What others will designate as the "preexistent germs" are not, within a Leibnizian context, *material ultimates* but secondary manifestations of a teleologically directed force that constitutes the deeper substance of beings. For this reason the system does not reduce to material mechanism. Material bodies at any level are only secondary phenomenal manifestations of this underlying dynamism.

Leibnizianism among the French: Du Châtelet's synthesis

A summary of the interpretations of Leibnizianism by his disciple Christian Wolff in his massive textbook treatment of Leibnizianism in the 1720, '30s and '40s is not attempted here except to acknowledge that Wolff systematized, reinterpreted, and in some subtle ways changed some aspects of Leibnizian theory (Hettche 2014). For this reason, it is more correct to see the background of my discussion as Wolffianism rather than the specific doctrines of Leibniz himself. These Wolffian interpretations will be encountered as they were refracted through Émilie Du Châtelet's *Institutions*.

Du Châtelet's remarkable, and for the French deeply influential, synthesis of Wolffian and Newtonian natural philosophy formed a contribution to what J. B. Shank has termed the "Newton Wars." Although aspects of the Leibnizian–Wolffian philosophy had been circulating in France before the publication of her work, associated with the *vis viva* controversy (Shank 2008, Chapter 7), Du Châtelet's widely reviewed work drew new attention to Leibnizian–Wolffian metaphysics and made her work a major vehicle by which the French actually absorbed the Leibnizian–Wolffian philosophy in dialogue with Newtonianism. Du Châtelet rendered the key points of this philosophical program easy to comprehend, with a concrete application to issues of mechanics. As Julien Offray de La Mettrie was later to comment in his *Traité de l'Âme*, "all the world knows of the monads since the brilliant recruitment [*acquisition*] the Leibnizians have made of M. du Chattelet [*sic*]" (La Mettrie 1751, I, 256). Commentary has typically focused on the importance of her work for discussions of physics and mechanics. Here I will develop on her reflections bearing on fundamental issues in the relation of living phenomena to matter that will, I claim, illuminate otherwise puzzling views that Buffon subsequently was to articulate.

The nine prefatory chapters to Du Châtelet's text constitute a separate treatise on Leibnizian–Wolffian metaphysics, matter theory, and epistemology. In this context she also raised consideration of the hotly debated issue of embryological development and the reigning preexistence solution to it in a discussion of the impossibility of metaphysical atomism.

In her critique she argues that material atoms cannot be the final terminus of reality. Rather, these must be underlain by "simple beings" (*êtres simples*) – her translation of the Leibnizian monads as this was refracted through Wolff's *Ontologia* (Hettche 2014, Section 8.1). As developed extensively in Chapter 9 of her work, the principle of sufficient reason undermines any ultimate material atomism. The ultimate "simple beings" are divine immaterial creations with teleological directedness constituted by perpetually acting *force*:

> The principle containing the sufficient reason for the actuality of an action, whatever it may be, is called *force*; because the simple power or faculty of action is only in the [simple] beings a possibility of action or passion, for which it is necessary [to give] a sufficient reason for its actuality. It is thus that one can say that an animal has the faculty of movement.
>
> (Du Châtelet 1740, VII, §§126, 137)

This *force* is the essence of bodies. As Du Châtelet summarizes it:

> One sees by that, that the true substances, that is to say, the simple beings, are active, since they bear in themselves the principle of their changes, that is to say, this force which is essential to them, which never leaves them and which cannot be extinguished.
>
> (Ibid., §§128, 139)

Thus an organism, as well as any body composed of material, would be ultimately underlain by dynamic teleological force:

> It is thus necessary to admit between this possibility of a sufficient reason for actuality, that is, a force which puts into action this power of action of the [simple] being … all that which happens in the composite [being] is found in the end in the simple beings; it follows that the simple beings have this force, which consists in a continuous tendency to action, and this tendency always has its effect when there is no sufficient reason which can prevent this action, that is, when there is no point of resistance.
>
> (Ibid., §§126, 137–138)

But there is, nonetheless, as we have seen earlier with Leibniz, a second phenomenal state of *matter* deriving from that of the *êtres simples*. For Du Châtelet, this material level also now serves as a practical terminus of the composition of things, a level of empirical rather than metaphysical inquiry. As she writes:

> One has seen above that the atoms or indivisible parts of matter are inadmissible, when they are regarded as simple, irresolvable, and primitive matter, because one cannot then give a sufficient reason for their existence; but when one recognizes that they draw their origin from simple beings, one can very well admit them. Because it is very possible, and experiments render it

> very probable, that there is in the Universe a certain determinate number of parts of matter that Nature never reduces into its principle [*ne résout jamais dans leur principe*], [and] which remain undivided in the present constitution of this Universe, and that all the bodies which compose it result from the composition and mixture of these solid particles; so that one can regard them as Elements endowed with figures, and with internal differences, which result from their parts. That Nature stops in the analysis of Matter at a certain fixed and determinate degree is rather probable, by means of the uniformity which reigns in its work, and through an infinity of experience [*d'expérience*].
>
> (Ibid., IX: §§172, 185–186)

This combination of a metaphysical argument for the impossibility of an ultimate material atomism, combined with the admission of a practical terminus of the division of entities in material particles with specifying sizes, shapes, and *internal* differences, allows empirical evidence to bear on the claimed existence of these dynamic *êtres simples* (ibid., §§172, 188). This theory supplies the foundation for the preexistent *germes* as the practical material termini of empirical division in organic beings, even though these are not the ultimate basis of organisms. Recent microscopic observations substantiate this:

> This irresolvability of the first Bodies becomes indispensibly necessary, if one adopts the system of germs [*sistême des germes*], which the new discoveries made by the microscopes seem to demonstrate. Everyone knows of those of M. Hartsoeker, and it becomes every day more probable that Nature acts only by development [*développement*]; now if each grain of wheat [*bled*] contains the germ of all the wheat it must produce, it is necessary that the actual divisions of Matter have limits, although these limits would be unassignable by us. It is thus very probable that there are particles of Matter of a certain determinate minuteness that Nature divides no further.
>
> (Ibid., 188)[5]

This gives us an odd "materialism" in practice, if not in theory. First, it is a *dynamic* "phenomenal" materialism, since the underlying *substance* giving rise to these particles is teleologically directed force and not extension, shape, or density. Second, these *material ultimates* are not homogeneous entities, and in this they differ from classical atoms. They are instead still specific "all the way down" to the monadic centers that underlie them.[6] This principle also underlies the fixity of species: "plants, animals [and] fossils all in the end each constantly produces its likeness with the attributes which constitute its essence" (ibid.). We can follow the implications of these remarks into the writings of her contemporary Buffon with some important illumination of the origins of his doctrines now made possible.

Buffon's organic molecules

The appearance of Du Châtelet's synthesis of issues in an accessible and semi-popular form in late 1740 placed it exactly in the time window when Buffon was in the process of moving from an early professional identity as a mathematician and member of the "Newtonian" party at the Académie des Sciences with an appointment of *adjoint-mechanicien*, to his powerful position as *intendant* of the Jardin du Roi, a position to which he was appointed in July of 1739 to replace Charles Du Fay, who had died suddenly that summer. This shift launched Buffon into his four-decade career that made him a dominant figure of the age in cosmology, anthropology, and natural history.

Buffon's explicit concern with issues related to the generation theory and biological questions following his appointment to this new position can be documented from 1741, when he communicated a letter on the topic to the Royal Society of London concerning the discovery of the regenerative powers of Abraham Trembley's famous "polyp" (Anon. 1742, 219n). It is reasonable to presume that this date also marked the beginnings of his theoretical and experimental inquiry into the generation issues that he pursued later in detail in discussions with his friend Pierre de Maupertuis, and in microscopic work with John Turbeville Needham (Sloan 1992b).

Although it has been common to place Buffon in a French "Newtonian" tradition (Hanks 1966, Chapter 2; Cassini 1992), a finer-grain discussion of the meaning of this designator suggests that he does not fit easily into this category (Wolfe 2014).[7] It is the claim of this chapter that the roots of his analysis of the relation of life to matter or the origins of his organic molecules theory lie elsewhere, and reflect his awareness of the work of Émilie Du Châtelet and his own reading of the *Institutions de Physique* itself. The hybrid character of Buffon's views themselves reflect the kind of synthesis of Newtonianism and Leibnizianism expounded by Du Châtelet in the *Institutions*. At the same time, Buffon will eliminate some of the critical metaphysical components of that synthesis.

Buffon can be documented to have been in residence in Paris during much of the period between August and December of 1739, the period when Du Châtelet was engaged in revisions of the *Institutions* and meeting with Parisian scientific figures. One dinner meeting they both attended is documented.[8] Other opportunities for interaction occurred in 1740 when Buffon was in Paris and Du Châtelet was often there to meet with her publisher on the final revisions of her text. Beyond this, there is little to document personal meetings. But, whatever may have been their personal contacts, shortly after the publication of the *Institutions* in December of 1740, Buffon had read, and was obviously impressed by, this novel work. As Claude Helvétius wrote to Du Châtelet, probably in January of 1741,

> The first use I have made of my new life [in Paris] is to write to you and tell you that I have seen M. de Buffon. He thinks and speaks of what he owes to your work, [and] he found it written with clarity, order, neatness, and

> precision in its language and ideas, and finally that he finds admirable all that is yours in this work.
>
> (Helvétius to Du Châtelet, *c.* January 1741 in Besterman 1958 II, Letter 255, 36)[9]

Exactly what Buffon may have taken from this early 1741 reading is difficult to document with precision, and this letter may express his hesitation to accept what he may have interpreted to be Koenig's Wolffianism. In the absence of illuminating background manuscript sources or direct textual references in Buffon's texts, it requires an exegesis of his printed texts themselves to penetrate where these connections may be found.[10]

As we look more closely at the matter–life relationship developed by Buffon in light of the discussions in the *Institutions*, several illuminating connections appear. As dated in the opening volumes of the published *Histoire Naturelle* (1749), Buffon completed the theoretical discussion of his *Traité sur la Génération* of the second volume of the *Histoire Naturelle*, expounding this theory of the *molécules organiques* and *moules intérieurs*, on February 9, 1746. It is then after this date that he embarked upon a set of microscopic experiments carried out in collaboration with John Turberville Needham and Louis Daubenton in 1747 and 1748 (Sloan 1992b).

In the 1746 "theoretical" discourse, Buffon first introduced his distinction between "living" and "dead" matter, claiming that "life and animation, in place of being a metaphysical degree of beings, is a physical property of matter" (*HN* II, Chapter 1, *OP*, 238 a–b). But exactly what this entails is in need of clarification. It does not imply an endorsement of a form of vital organic *atomism*. His initial distinction is between "inert" (*brut*) and "living" (*vivant*) matter, the first being "inactive, without sense, and only constrained by the laws of mechanics, obeying only the force generally spread throughout the universe" (ibid., 234 b). Animate beings (*l'animal*), by contrast, are coordinated compounds that "join together all the powers of Nature, [and] the forces which animate it are specific and particular…" (ibid.).

A second point is essential to observe. It seems evident that Buffon is accepting at this time the theory of the preexistent *germes* similar to the form endorsed by Du Châtelet (Du Châtelet 1740, IX, 172). This means that the *germes* are organized derivatives of a more fundamental substrate. This concept is used by Buffon to explain the phenomenon of regeneration in Trembley's hydroid, in which the regenerating organism is seen as a composite of such preformed germs, each of which is capable of developing into the whole (*HN* II, *OP*, 238b). As with Du Châtelet, the *practical* terminus of analysis is in the preexistent *germes*. But, unlike her view, these are themselves seen as underlain by more primitive and living material "organic parts" rather than immaterial force:

> This leads us to believe that there is in Nature an infinity of existing and living organic particles [*de parties organiques*], of which the substance [*substance*] is the same as that of the organized beings, just as there is an

> infinity of dead particles [*particules bruts*] similar to the dead bodies [*corps bruts*] that we are acquainted with. And as it would perhaps be necessary for thousands of small cubes of salt to accumulate in order to make the sensible individual of a grain of sea salt, it is also necessary for thousands of organic particles, similar to the whole, to form a single one of the germs [*germes*] which contains the individual of an elm or a polyp.
>
> (Ibid.)

One implication of this is that we do not find in Buffon a theory of *undifferentiated* material organic particles that are simply structured in specific ways by a form-giving "Newtonian" internal mold. The differentiation of specific properties instead, as we have seen Du Châtelet argue, is ultimate. Speaking of the results of microscopic observations by Leeuwenhoek and others that predate his own later studies with John Turbeville Needham later, he comments:

> The animals and plants which can multiply and reproduce themselves by all their parts, are organized bodies [*corps organisez*] composed of other similar organic bodies, of which we visually discern the accumulated quantity, but of which we are not able to perceive the primitive parts except by reason and the analogy we come to establish.
>
> (Ibid., 239a)

Read against the statements of Du Châtelet in Chapter 9 of the *Institutions*, we see some striking similarities, and also important differences. With Du Châtelet, Buffon recognizes a practical terminus of the division of matter in the form of material bodies that have *internal differences* and which in conglomeration constitute the *germes*. The *germes* are derivative and secondary, rather than ultimate. The organism can thus be analyzed into a subordinate collection of smaller, but *specifically differentiated* organic entities, just as a cube of salt can be seen as composed of lesser cubes of salt:

> And just as it is necessary to separate, pulverize, and dissolve a cube of sea-salt in order to perceive, by means of crystallization, the small cubes of which it is composed, it is similarly necessary to separate the parts of an elm or polyp in order to then recognize, by means of planting or expansion [*développement*], the small elms and the small polyps contained in these parts.
>
> (Ibid., 239a–b)

This establishes for Buffon a hierarchy of levels between a more elementary foundation, possibly implying what he may mean by *substance*, and the material that comprises visible organisms and other bodies. At this most elementary level are *dynamic* entities, similar in some respects to Du Châtelet's *êtres simples*, but with an important difference we will explore. These are specifically determinate, and in combination are manifest in the *germes* which presumably enter the domain of the microscopically visible. These *germes* are also *miniatures of the whole phenomenal organism.*

Another "Leibnizian" strand concerns the inherent activity Buffon gives to the concept of Nature. As Du Châtelet had defined "Nature":

> When one speaks of Nature in general, one understands an internal principle of changes which occur in the world: this isn't a lesser God, distinct from the world, which would govern this machine; it isn't a motive force joined to other properties which with it composes the essence of the body. This motive force is the only principle of motion in the Universe, and it is by means of this that one can comprehend why the possible changes become actual.
>
> (Du Châtelet 1740, VIII: §§161, 174)

Buffon embraces views similar to these comments, and adopts much of the teleological language we find in Du Châtelet: "Nature does not tend to the lifeless [*du brut*], but the organic [*organique*], and when it does not reach this goal [*but*], it is only because there are resistances [of inert matter] [*des inconvéniens*] which oppose it" (*HN* II, *OP*, 245a). Elsewhere he comments:

> Nature in general appears to me to tend much more to life [*la vie*] than to death [*mort*]. It seems that it seeks to organize bodies as much as possible, [and] the multiplication of the germs [*des germes*], that can be multiplied almost infinitely, is one proof of this.
>
> (Ibid., 244b)

This vital Nature is the deeper source of the *moules intérieurs*:

> There exists in Nature some forces, such as that of gravity, which are relative to the interior of matter, and which have no relation with the exterior qualities of bodies, but which are active on the most intimate parts, and which penetrate them at all points; these forces, as we have proven, never come under our senses, because their action is made on the interior of bodies.
>
> (Ibid., 247 a–b)

This is not a Newtonian claim. The interior activity issuing from Nature is neither simply attraction nor a similar Newtonian microforce. It is also inherently vitalizing and dynamic. It organizes the *parties organiques* and it accounts for nutrition and development (ibid., 247b).[11] In company with the specificity of these underlying organic particles, Nature, rather than matter, ultimately accounts for vitality. In summary, the ingredients we have identified for focus to this point – matter, *germes*, Nature, and the fundamental dynamism underlying bodies that for Buffon distinguishes "life" from "nonlife" – all seem to have roots that lead us to the probable importance of Du Châtelet's synthesis for a deeper understanding of Buffon's discussion of the *molécules* and *moules*.

But there are also fundamental differences between Du Châtelet and Buffon that prevent any simple location of Buffon in the Leibnizian–Wolffian tradition articulated by Du Châtelet. One difference concerns the relation of *substance* to these elementary *organic parts*. For Du Châtelet, following Leibniz and Wolff, the distinction of *êtres simples* from *germes* and other phenomenal compound material beings is a crucial distinction. Du Châtelet had indeed allowed that for practical purposes, the issues do not need to be pressed to this metaphysical level of analysis. As we have seen above, it is generally sufficient to remain at the level of phenomenal matter for our scientific purposes. But there is no conflation of these ultimate material particles and *êtres simples*.

Buffon makes no such clear distinction of *matière* and *substance*, and instead simply makes matter *itself* alive at the most fundamental level of the *molecules organiques* that lie behind the *germes*. He simply amputates the more metaphysical level of inquiry that underpinned the Leibnizian distinctions of matter and substance. Such metaphysical knowledge is rendered epistemologically unavailable by Buffon, rather than simply a practical difficulty: "In Nature … the abstract does not exist; nothing is simple and all is compound; we never penetrate to the intimate structure of things" (*HN* II, *OP*, 240a). Except for appeals to a substantive and vital Nature that gives the ultimate source of this dynamism, there is no deeper appeal by Buffon to an underlying force-dynamics that would connect us more closely with the Leibniz–Wolffian program.

A second fundamental departure from Du Châtelet's Leibnizianism is more complicated. Although Buffon uses, as we have seen, strongly teleological language in describing the activity of Nature in general, Buffon also rejects particular explanations in terms of final causes, meaning by implication such Leibnizian principles as that of sufficient reason appealed to heavily in Du Châtelet's analysis of the force–substance–matter distinctions. Such questions as to why is matter impenetrable and what is gravity, motion, or rest? – are deemed unanswerable. To such questions, "one can only respond by the question itself: it is such, because in fact it is such" (ibid., 242a).

Conclusion

In this detailed investigation of one strand of the broader "vitalist" revolution of the mid-Enlightenment, we have followed how a complex metaphysical program developed by Leibniz and Wolff, and then expounded for a French reading public by Émilie Du Châtelet, led into the formulation of an "active" materialism that made vital powers a fundamental dimension of matter.

Du Châtelet's synthesis retained, as we have seen, the fundamental metaphysics of Leibnizianism, and avoided an "ontological" vital materialism. But she also opened a door by which others – Buffon only partially – could draw from her writings the grounds for a more comprehensive vital materialism. La Mettrie, who directly engaged Du Châtelet's philosophical program in his *Histoire Naturelle de l'Ame* of 1745 (La Mettrie 1751), simply cut this "vitalization" of matter more completely loose from its original metaphysical moorings in the

Leibnizian philosophy, and took from Du Châtelet primarily the authorization to grant a fundamental "motive power" to matter – *sentiment* – which could also serve as the basis of feeling and consciousness itself. This development is from a metaphysics of substance to an empirically based philosophy of Nature with living beings constituted by a vital conception of matter. In this we can see emerge one important ingredient of what will later be termed a separate science, devoted to analyzing the principles of living matter.

Acknowledgments

I wish to thank Charles Wolfe, Cécilia Bognon, Monica Solomon, and M. Katherine Tillman for valuable comments on this manuscript. I am indebted to the stimulation provided by the Notre Dame Du Châtelet seminar at Notre Dame in 2016 directed by Professor Katherine Brading, and to discussions in the Notre Dame Du Châtelet conference on the *Institutions de Physique*, April 26–28, 2018.

Notes

1 On the incorporation of her work in the articles on Leibnizianism in the *Encyclopédie* formally attributed to Samuel Formey, see Maglo 2008. I am also indebted to the new research on Formey's incorporation of Du Châtelet's *Institutions* in his *Encyclopédie* articles as presented in her unpublished paper (Seul 2018). Cited by permission.
2 See below, note 3.
3 This third variant of the preexistence theory, not always distinguished by other scholars, only entailed the preexistence of certain rudimentary internal relations of parts in these *germes*, and did not imply the preexistence of miniature and fully formed organisms encased within one another. These germs are widely disseminated in nature rather than encased in the ovary, testes, or plant seeds and are taken in with food and developed under proper conditions. Such "germs" also retained some aspects of Aristotelian "potentiality" in this concept. On this concept, see discussion in Roger 1994, Pt. II, Chapter 3 and Sloan 2002.
4 For discussion of the complexities in this see Phemister 2011.
5 This passage is radically altered in the second edition of 1742, and drops reference to the preformed *germes* (Du Châtelet 1742). Illuminating the reasons for this alteration take us beyond this chapter.
6 The relationship of Du Châtelet's thesis to the "naturalization" of the Leibnizian monad, which Georges Canguilhem has seen as the background of the cell theory, is an issue that would require separate discussion (Canguilhem 2008, Appendix 2). Du Châtelet's clear distinction between the metaphysics of substance and the practical terminus of empirical enquiry at the level of microscopic material entities, would not allow an easy identification of matter and substance unless the larger metaphysical edifice were to be stripped away. This conflation is, however, exactly that being made by Buffon and others identified by Canguilhem. I thank Charles Wolfe for valuable comments on this point.
7 Charles Wolfe's important discriminations of the variants of Newtonianism adopted by life scientists of the period assist us in refining Buffon's location in relation to the Newtonian tradition. I concur with Wolfe that Buffon falls between camps in some of the broader positions within the French Leibniz–Newton debate characterized by Shank (2008, Chapter 7), and he does not easily fit into "experimental" Newtonianism as it has been characterized by Zammito (2018, Chapter 2). Epistemologically, his

development from an early Newtonian "positivism," expressed in his preface to his translation of Stephen Hales's *Vegetable Staticks* in 1735, to a more assertive conception of the certitude of the "concrete' over the "abstract" sciences that commences with the *Premier Discours* to the first volume of the *Histoire Naturelle* in 1749, moved him distinctly away from the characteristic epistemological feature of Newtonianism and more distinctly in the direction of the rationalism of Leibnizianism. This is not to claim that Buffon is a "Leibnizian" in any orthodox sense.

8 In August of 1739, Du Châtelet and Buffon were at a dinner at the residence of Mme Françoise d'Issembourg de Graffigny that also included Dortous de Mairan, Réné Réaumur, and Bernard de Fontenelle (Vaillot 1988, 124). I have determined the dates of Buffon's presence in Paris through the record of his attendances at the meetings of the Académie des Sciences (*Procès verbaux de l'Académie des Sciences*, Archives de l'Académie des Sciences, Paris). This evidence suggests a dating of the quoted Helvétius letter around January 1741. Du Châtelet's *Institutions* appeared in December of 1740. For further details, see the list of known correspondence in this period in Hanks 1966, Appendix.

9 Le premier usage que je fais de ma nouvelle vie est de vous écrire et de vous mander que j'ai vu M. de Buffon. Il pense et parle comme il le doit de votre ouvrage, il le trouve écrit avec clarté, ordre, netteté, précision dans les mots et les idées, enfin il trouve admirable tout ce qui est à vous dans votre ouvrage.

(As in Besterman 1958, II, 36)

10 Buffon intentionally destroyed nearly all his manuscripts, and there is no surviving correspondence or other material from this period able to illuminate this interaction further. An electronic name search through Buffon's *Œuvres Complètes* and *Correspondance* (Buffon, www.buffon.cnrs.fr) discloses no mention of Du Châtelet or of the *Institutions*. But a similar search reveals no references to Locke, Spinoza, Wolff, or Condillac, only eight to Leibniz, mainly related to his cosmology, and 17 to Newton, mainly connected with his theory of the planets. One suggestive point of connection with Du Châtelet is found in Buffon's endorsement of the creative role of hypotheses in science. This use of "hypotheses," condemned in the rhetoric of the French Newtonians as "the poison of reason and the plague of philosophy" (Du Châtelet 1740, 75), was explicitly defended by Du Châtelet in Chapter 4 of her *Institutions* as playing a necessary role in science. Buffon similarly endorses the creative role of imaginative hypotheses in his discussion of life and matter and organic generation theory in the crucial discourse that opens Volume 2 of the *Histoire Naturelle*, where he appeals to the fertility of "hypotheses" in formulating his theory of the *moules intérieurs* (Buffon "Histoire générale des animaux," *Histoire Naturelle* II (1749) as in Piveteau 1954 (1749), (*OP*) 243a–b. See also Sloan 1992a). No mention of Du Châtelet's work is made in the critical apparatus for the *Premièr Discours* in the critical edition of the *Histoire Naturelle* by Stéphane Schmitt and Cédric Crémière in Buffon (2008). For ease of reference I will cite the Piveteau collection of texts extracted from the Imprimerie Royale edition of the *Histoire Naturelle (OP)*. Translations are my own.

11 Buffon only deals in more detail with this notion of a "vitalizing" nature in the "De la nature: Premier vue" of 1764. There he speaks of Nature as:

> The system of laws established by the Creator, for the existence of things and for the succession of beings. Nature isn't a thing, because this thing would be all; Nature isn't a being [*être*], because this being would be God; but one can consider it as an immense living power, which embraces all, which animates all, and which is subordinate to the primary Being.

(Buffon HN XI [1764], *OP*, 31a)

References

Alexander, H. G. (ed.). 1984. *The Leibniz-Clarke Correspondence*. Manchester: Manchester University Press.

Anon. 1742. Extract of a letter from J. F. Gronovius, M. D. at Leyden, November 1742 to Peter Collinson, F. R. S. concerning a water insect, which, being cut into several pieces, becomes so many perfect animals. *Philosophical Transactions of the Royal Society of London, 42*: 218–220.

Barber, W. H. 1955. *Leibniz in France from Arnauld to Voltaire: A Study in French Reactions to Leibnizianism, 1670–1760*. Oxford: Clarendon.

Barber, W. H. 1967. Mme Du Châtelet and Leibnizianism: the genesis of the *Institutions de Physique*. In. W. H. Barber *et al.* (eds.), *The Age of Enlightenment* (pp. 200–222). Edinburgh: Oliver and Boyd.

Besterman, T. (ed.). 1958. *Les lettres de la marquise Du Châtelet*. Geneva: Institut et Musée Voltaire.

Brading, K. and A. Seul. 2015. Reading the *Institutions de Physique* as a philosophical text: content and context in relation to the *Encyclopédie*. Presented at Oxford University Conference, May 14, 2015. Cited by permission.

Buffon, G.-L. 1954. *Œuvres philosophiques*, J. Piveteau (ed.). Paris: Presses Universitaires de France.

Buffon, G.-L. 2008. *Œuvres completes*, Vol. 2, Stéphane Schmitt and Cédric Crémière (eds.). Paris: Honoré Champion.

Buffon, G.-L. 2013. *Buffon et l'histoire naturelle: l'Edition en ligne*, Pietro Corsi and Thierry Hoquet (eds.). Electronically available at www.buffon.cnrs.fr.

Canguilhem, G. 2008. *Knowledge of Life*, P. Marrati and T. Meyers (eds.), S. Geroulanos and D. Ginsburg (trans.). New York: Fordham University Press.

Cassini, P. 1992. Buffon et Newton. In J. Gayon *et al.* (eds.), *Buffon 88* (pp. 299–308). Paris: Vrin.

Detlefsen, K. 2014. Émilie Du Châtelet. In E. N. Zalta (ed.), *The Stanford Encyclopedia of Philosophy*. Electronically available at http://plato.stanford.edu/archives/sum2014/entries/emilie-du-chatelet.

Diderot, D. 1956. *Rameau's Nephew and Other Works*, J. Barzun and R. Bowen (trans.). Indianapolis, IN: Bobbs-Merrill.

Distelzweig, P., B. Goldberg, and E. Ragland (eds.). 2016. *Early Modern Medicine and Natural Philosophy*. Dordrecht and New York: Springer.

Du Châtelet, G. 1740. *Institutions de physique*. Paris: Prault. Electronically available at http://gallica.bnf.fr/Bibliothéque nationale de France.

Du Châtelet, G. 1742. *Institutions de physique de Madame la Marquise Du Chastelet, nouvelle edition, corrigée & augmentée considerablement par l'auteur*. Amsterdam. Electronically available at https://books.google.com/books/about/Institutions_Physiques_V1_1742.html.

Duchesneau, F. 2010. *Leibniz: Le vivant et l'organisme*. Paris: Vrin.

Duchesneau, F. 2011. Leibniz versus Stahl on the way machines of nature operate. In J. E. H. Smith and O. Nachtomy (eds.), *Machines of Nature and Corporeal Substances in Leibniz* (pp. 11–28). Dordrecht: Springer.

Gaukroger, S. 2010. *The Collapse of Mechanism and the Rise of Sensibility*. Oxford: Oxford University Press.

Gerhardt. C. J. (ed.). 1875–90. *Die philosophischen Schriften von Gottfried Wilhelm Leibniz*, 7 Vols. Berlin: Weidman. Reprint Hildesheim: Olms 1960–1961.

Hanks, L. 1966. *Buffon avant l'histoire naturelle*. Paris: Presses Universitaires de France.

Hettche, M. 2014. Christian Wolff. In E. N. Zalta (ed.), *The Stanford Encyclopedia of Philosophy*. Electronically available at http://plato.stanford.edu/archives/sum2014/entries/Christian-Wolff.

Iltis, C. 1977. Madame Du Châtelet's metaphysics and mechanics. *Studies in History and Philosophy of Science Part A*, *8*: 29–48.

Janik, L. G. 1982. Searching for the metaphysics of science: the structure and composition of Madame Du Châtelet's *Institutions de physique*, 1737–1740. *Studies on Voltaire and the Eighteenth Century*, *201*: 85–113.

La Mettrie, J. O. de. 1751, 1987. *Œuvres philosophiques*. Vol. 1. London: Nourse. Reprinted Paris: Fayard.

Maglo, K. 2008. Mme Du Châtelet, l'*Encyclopédie*, et la philosophie des sciences. In Ulla Kölving and Olivier Courcelle (eds.), *Émilie Du Châtelet: Eclairages & documents nouveaux* (pp. 255–266). Ferney: Centre International d'Étude du XVIIIe Siècle.

Nachtomy, O., and J. E. H. Smith (eds.). 2014. *The Life Sciences in Early Modern Philosophy*. Oxford: Oxford University Press.

Perrault, C. 1721. La mechanique des animaux. In *Œuvres diverses de physique et de mechanique*, 2 Vols. Leiden: P. van der Aa.

Phemister, P. 2005. *Leibniz and the Natural World: Activity, Passivity and Corporeal Substances in Leibniz's Philosophy*. Dordrecht: Springer.

Phemister, P. 2011. Monads and machines. In J. E. H. Smith and O. Nachtomy (eds.), *Machines of Nature and Corporeal Substances* (pp. 39–60). Dordrecht and New York: Springer.

Pinto-Correia, C. 1997. *The Ovary of Eve: Egg, and Sperm, and Preformation*. Chicago, IL: University of Chicago Press.

Piveteau, J. (ed.). 1954. *Buffon: Œuvres philosophiques*. Paris: Presses Universitaires de France.

Reill, P. H. 2005. *Vitalizing Nature in the Enlightenment*. Berkeley, CA: University of California Press.

Roe, S. A. 1981. *Matter, Life, and Generation: Eighteenth-Century Embryology and the Haller-Wolff Debate*. Cambridge: Cambridge University Press.

Roger, J. 1994. *Les sciences de la vie dans la pensée française du dix-huitième siècle*, 3rd edition. Paris: Albin Michel.

Seul, A. 2018. *Du Châtelet and the Encyclopedia of Diderot and d'Alembert*. Notre Dame Du Châtelet Conference, April 28.

Shank, J. B. 2008. *The Newton Wars and the Beginning of the French Enlightenment*. Chicago, IL: University of Chicago Press.

Sloan, P. R. 1992a. L'hypothétisme de Buffon: sa place dans la philosophie des sciences du dix-huitième siècle. In J. Gayon *et al.* (eds.), *Buffon 88* (pp. 207–221). Paris: Vrin.

Sloan, P. R. 1992b. Organic molecules revisited. In J. Gayon *et al.* (eds.), *Buffon 88* (pp. 415–438). Paris: Vrin.

Sloan, P. R. 2002. Preforming the categories: eighteenth century generation theory and the biological roots of Kant's *a-priori*. *Journal of the History of Philosophy*, *40*: 229–253.

Smith, J. E. H. (ed.). 2006. *The Problem of Animal Generation in Early Modern Philosophy*. Cambridge: Cambridge University Press.

Smith, J. E. H. 2011. *Divine Machines. Leibniz and the Sciences of Life*. Princeton, NJ: Princeton University Press.

Smith, J. E. H., and O. Nachtomy (eds.). 2011. *Machines of Nature and Corporeal Substances in Leibniz*. Dordrecht: Springer.

Vaillot, R. 1988. *Avec Madame Du Châtelet, 1734–1749*. Oxford: Taylor Foundation.

Wolfe, C. T. 2011. Why was there no controversy over life in the Scientific Revolution? In M. Dascal and V. D. Boantza (eds.), *Controversies within the Scientific Revolution* (pp. 187–219). Amsterdam: John Benjamins Press.

Wolfe, C. T. 2014. On the role of Newtonian analogies in eighteenth-century life science: vitalism and provisionally inexplicable explicative devices. In Z. Biener and E. Schliesser (eds.), *Newton and Empiricism* (pp. 223–261). Oxford: Oxford University Press.

Zammito, J. 2018. *The Gestation of German Biology: Philosophy and Physiology from Stahl to Schelling*. Chicago, IL: University of Chicago Press.

4 The philosophical uptake of Caspar Friedrich Wolff in German philosophy after 1770

Tetens, Herder, Kant and Blumenbach

John H. Zammito

Introduction

In the third quarter of the eighteenth century, life science in Germany had not yet calved away from *Naturlehre* (the "physical" sciences, generally), or even from traditional natural philosophy, into an independent special science. It remained part of the disciplinary matrix of philosophy (*Weltweisheit*) in general. Moreover, the issues it posed regarding organism, instinct, mind–body relations, generation etc. had enormous bearing upon central questions in Western philosophy from its Greek origins to contemporary elaborations in the wake of Descartes and the other moderns (Wright and Potter eds. 2000; Niewöhner and Seban eds. 2001; Smith ed., 2006). For Germans, this meant that the school-philosophical tradition of Christian Wolff – and its critics, as well – had to take cognizance of these developments. One of the decisive issues in this philosophical engagement with biology before the full establishment of that special science was the question of epigenesis, especially as it became controversial in the exchange between the dominant figure in German life science, Albrecht von Haller (1708–1777), and the newcomer Caspar Friedrich Wolff (1734–1794) (Roe 1981). This chapter will consider the reception of Wolff, specifically by some of the most important philosophical commentators of the later eighteenth century in Germany.

Caspar Friedrich Wolff and epigenesis

A great deal remains to be done in accurately placing Caspar Friedrich Wolff in the history of life science in the eighteenth century. Without question, Wolff elaborated the most important experimental and theoretical case for epigenesis in the mid-eighteenth century. He conceived *vis essentialis* as a Newtonian force that induced, through certain chemical processes, the production of organic matter out of inorganic matter in regular and empirically demonstrable patterns (Wolff 1759, 1764, 1774; Roe 1979, 1981; see also Herrlinger 1959; Aulie 1961; Lukina 1975; Mocek 1995; Gaissinovich 1990; Breidbach 1995; Duchesneau 2006).

Modern usage regarding epigenesis set out from William Harvey's 1651 text, *On Generation*, which defined it as the characteristic of an organism that "all its parts are not fashioned simultaneously, but emerge in their due succession and order.... For the formative faculty ... acquires and prepares its own material for itself" (Harvey 1651, 366). Harvey's eighteenth-century successors, Pierre Moreau de Maupertuis and Georges LeClerc de Buffon, believed all the resources for life lay *immanent* in nature. Epigenesis was a theory of *materialism*, however "vital" (Reill 2005). But scholars have aptly noted their hesitancy regarding blind mechanism. They both held out the requirement for some design: Buffon's *moule intérieur*; Maupertuis's metaphor of knowledge or desire at the particle level (Buffon 1749; Maupertuis 1756). Buffon's *moule intérieur* was a reformulation of Harvey's "formative faculty," a principle of design that set in motion determinate mechanisms of organic development in the embryo. Some question whether Buffon should even be considered an epigenesist, since there was not a little "preformation" in his theory, but *he* clearly distanced himself from preformation.[1] Moreover, in his time he was certainly lambasted as a materialist, an "Epicurean" (Hoquet 2005). Consequently, there are grounds for taking him for an epigenesist. Buffon invoked an analogy between his "microforce" and Newton's characterization of gravity. That is, for experimental Newtonianism *effects* were empirically demonstrable even if the *cause* remained obscure. This became a consistent methodological premise among all subsequent theorists of epigenesis in the eighteenth century (Gaissinovich 1968; Hall 1968; Wolfe 2014).

How are we to account, in any event, for the epigenesis of Caspar Friedrich Wolff? *Theoria Generationis* (1759) and its defensive elaboration in *Theorie von der Generation* (1764) proved unquestionably epochal. Moreover, Jean-Claude Dupont is clearly right that "the full originality of Wolff's theory of epigenesis undoubtedly appears better in *De formatione intestinorum* (1768–1769) than in *Theoria generationis*" (Dupont 2007, 43; Wolff 2003). Yet there is good evidence that – thanks to Albrecht von Haller's dismissal of Wolff's findings, echoed enthusiastically by Charles Bonnet and then maintained until Haller's death in 1777 by the only other serious theoretical rival of Wolff in German embryology, Johann Friedrich Blumenbach – Wolff became somewhat marginalized in German life science for a considerable span. Precisely in that span fell the publication of his masterpiece, *De Formatione Intestinorum*. Haller dismissed it, and Bonnet did not even bother to read it, given Haller's dismissal. The text did not really receive recognition until the early nineteenth century, with its translation into German by Johann Meckel the younger, and then its enthusiastic adoption by Karl von Baer and others, raising it to the status of a paradigmatic work in experimental life science (Wolff 1812; Baer 1828). It was rather the works of Haller and Bonnet that dominated consideration of the issue, until they were in turn overthrown by Blumenbach's notion of the *Bildungstrieb*. In the 1760s, Haller and Bonnet articulated a strong defense of preformation in response first to Maupertuis and Buffon and then, more fundamentally, to Wolff. In the course of their argument, they reformulated preformation in a direction that could accommodate some epigenetic elements (Sloan 2002). Günter Zöller characterizes their view as follows: "preformationism is primarily a theory concerning the generation of distinct parts (organs) in the growing embryo. It maintains that growth is *quantitative* growth of preexisting parts … no qualitative embryological growth or formation of new parts" (Zöller 1988, 79).

Blumenbach's *Bildungstrieb* essays (Blumenbach 1781, 1789) made light of Wolff's work. So did the crucial philosophical commentary on the whole debate over preformation and epigenesis by Johann Nicolaus Tetens (Tetens 1777). A talisman of the eclipse of Wolff in German life science was the surprise expressed by Goethe in the early nineteenth century over his discovery of Wolff's work – explicitly via the reception of Blumenbach by Kant (Goethe 1988). Wolff was not a major figure in the German discourse on vital force (*Lebenskraft*) in the last decades of the eighteenth century. A token of this is the absence of all reference to him in the widely cited text of Medicus (Medicus 1774).

Wolff made one final effort to reestablish leadership in the field of embryology by organizing a contest in the St. Petersburg Academy of Science in 1782 over the question of nutrition as a vital force. He induced Blumenbach to participate, and, as judge of the contest, recognized him as one of the two prizewinners (along with Carl Friedrich von Born). Then he confronted Blumenbach's view with a far longer and more detailed exposition of his own ideas and published all the texts together under the auspices of the St. Petersburg Academy in 1789 (Wolff 1789a, 1789b). This challenge to Blumenbach induced him to make some methodological revisions of his own position, as we shall note below (Blumenbach 1789). Even so, Wolff did not succeed in capturing a leading role in the German discussions. Instead, Blumenbach's reformulated *Bildlungstrieb* became the common argot of the decades after 1790 (see, e.g., Reil 1795; Schelling 1798). Only with Meckel's retrieval of Wolff's text in 1812 did Wolff's star rise again.

Wolff studied at the Collegium Medico-Chirugicum of Berlin, starting in 1753, above all with the anatomist Johann Friedrich Meckel the elder (1724–1774) and with the botanist Johann Gottlieb Gleditsch (1714–1786), the curator of the royal gardens in Berlin (Uschmann 1955). In 1754 he advanced to the University of Halle. After nine semesters, he defended his famous dissertation, *Theoria Generationis*, in November 1759 (Wolff 1759; 1774; 1896). How did he come to the topic and the approach of his dissertation? No one in the medical faculty of the late 1750s in Halle evidenced the boldness and vision that could have inspired the young Wolff to such a daring dissertation.[2] Anatomical and botanical resources were mediocre at best at Halle and yet these were the overweening preoccupations in Wolff's experimentally driven dissertation. The most likely director of studies for Wolff, as far as Ilse Jahn can establish, was Philipp Adolf Böhmer (1717–1789):

> Böhmer … announced for summer 1756 a public lecture [series] "*de homine generatione*" [on human generation], in winter 1757/58 led a private course on anatomical preparations and demonstrations, and in summer 1758 gave a lecture series on the *Institutiones physiologicas* of [Christian Gottlieb]

> Ludwig [1709–1773], whom Wolff explicitly named [in his dissertation] as a representative of epigenesis.
>
> (Jahn 1998/1999, 43)

But Böhmer was no innovator in theory or experiment.

Jahn has persuasively established that the inspiration for Wolff did not come from Halle at all, but rather from his earlier years in Berlin in the person of the president of the Prussian Academy, Pierre Moreau de Maupertuis. Jahn dwells on the quite specific "issues and … formulations of research programs concerning 'generation'" that appeared in Letter VII ("On the Generation of Animals") and in Letter XIX ("On the Progress of the Sciences") of Maupertuis's work of 1752, *Lettre sur le Progrès des Sciences*, which presented a "bounty of forward-looking research programs, questions and tasks to be undertaken" that can be quite specifically linked to what Wolff set about in his dissertation and thereafter (Jahn 1998/1999, 47–50; Maupertuis 1752). She writes:

> When one considers the daring and conviction with which Wolff went about his microscopic observations after the example of Buffon and Needham, and the originality with which he interpreted theoretically his results concerning the epigenetic development of plant germs and chicken embryos, whereby he introduced a new "theory of generation," one can certainly surmise that he carried out the commission of Maupertuis and simultaneously redeemed the latter's legacy to the Berlin Academicians. This impression grows yet stronger when one also takes into consideration the later works, [especially] the work on the formation of the intestine in the fertilized chick (Wolff 1766–1767) which was composed while he was still in Berlin, [and] which Karl Ernst von Baer characterized as "the greatest masterpiece the we know in the field of observational science."
>
> (Jahn 1998/1999, 50–51)

Particularly revealing is that Maupertuis celebrated the microscopic experiments of Buffon and Needham as setting the terms of investigation for a whole "new nature" (Jahn 1998/1999, 49–50). Haller and Bonnet were arguing at that same moment for the utter inadequacy of these experiments and their theoretical interpretation (Haller 1752, 1981; Bonnet 1776, 1782). That Wolff took up this specific experimental tradition suggests a positive inspiration that could not have come from any plausible source in Germany besides Maupertuis.

For evidence that Maupertuis was actively eliciting an endeavor along such lines in Germany, Jahn points to a remarkable prize competition announced by the Berlin Academy around 1755 for the section of experimental natural philosophy: "Whether all living things, as much in the animal kingdom as in the vegetal kingdom, arise from an egg fertilized by a germ, or from a prolific matter, analogous to the germ [*une matière prolifique, analogue au germe*]" (Jahn 1999, 98). Jahn notes that *no* answers were submitted, even when the competition was extended to 1759. A renewed call in 1761–1763 elicited four submissions but no prize was awarded. Quite simply, no one in the radius of the Berlin Academy wanted anything to do with this question: no one, that is, but Caspar Friedrich Wolff! Thus, crucially, we may recognize Wolff as a German life scientist directly and positively inspired by the innovations of French vital materialism. *Theorie von der Generation* not only inveighed against the preformationist notion, by which, it alleged, all organic bodies became blunt miracles, but also offered a clear sense of the *theoretical* alternative offered by *epigenesis*: "a nature that destroyed itself and that created itself again anew, in order to produce endless changes, and to appear again and again from a new side," "a living nature, which through its own forces produced endless changes" (Wolff 1764, 73).

Wolff's experimental observations concentrated on the fertilized chicken embryo, following upon the work of Malpighi (and, without realizing it, paralleling the concurrent research of Haller, published in 1758). In the opening section of Part II, on the development of animals, Wolff asserted that the origin of organic form in animals paralleled what he had established in his account of plant growth: "a mass which ... in general consists merely of a few loosely conglomerated little globes [*Kügelchen*] and simply amassed one upon another, transparent, moveable and almost fluid..." (Wolff 1896, 1). The cause of organic formation, just as with plants, was the "essential force [*vis essentialis*]," which induced movement of the nutritive fluid to all the parts of the organism – notably, at the outset, without need of a vessel system or a heart to pump it. He drew comparison to the work on the movement of lymphatic fluids by Meckel and Monro. In §242, Wolff enunciated his core thesis succinctly: "Thus the essential force, together with the solidification propensity of the nutritive fluid, constitutes a sufficient principle of all development both in plants and in animals" (Wolff 1896, 60). The crucial difference between animal organization and plant organization was that the solidification propensity in animal tissue worked far more slowly than in plants.

After a seemingly generous opening line in his 1760 review of Wolff's dissertation, Haller proceeded to browbeat his young rival in no uncertain terms over the balance of their correspondence.[3] He embodied an academic culture which spurned Wolff as an insouciant beginner who dared to challenge professional and professorial authorities in the medical-scientific world of mid-century Germany. Wolff had hoped for better from Haller. He really expected that Haller would find his work not only experimentally but theoretically congenial.[4] When Wolff began his investigations, he was not aware of Haller's fateful conversion to preformation. The major articulation of that conversion appeared only a year before Wolff defended his dissertation and probably before Wolff could adapt his experiments or their interpretation to incorporate Haller's views. Haller's *Sur la Formation du Coeur dans le Poulet* appeared in 1758, obviously in French (Haller 1758; Cherni 1998). How swiftly this became accessible to the research community in Berlin and Halle is not altogether clear. There is no mention of Haller's work on chicken embryos in Wolff's dissertation, though this was its central concern and Wolff did cite Malpighi on chicken embryos as well as many other works by Haller, so that it would seem that, had he known of Haller's monograph, it would have been referenced. To be sure, Wolff would never have countenanced preformation, even under the auspices of Haller; his experiments and their epigenetic interpretation remained incontrovertible, as he saw it, and as he would establish beyond question in his later work on the development of the intestine.

Wolff's *Theorie von der Generation* aimed not only to persuade the learned public of his epigenetic approach but also to (re)convert Haller. He wrote:

> I am confident that [Haller], when I have only had the opportunity to present my reasons, [based on a knowledge of animal physiology] with which he is quite familiar, will soon completely agree with me, and allow the argument for continuity [epigenesis] to go forward. *From Mr. Bonnet* I cannot hope the same. He seems to me, like many who take themselves for physiologists, to be quite far from such a knowledge of the nature of animals.
>
> (Wolff 1764, 117)

Wolff had little respect for Bonnet: "everything that Mr. Bonnet has written is for the most part derived from Mr. von Haller. Here and there he has mixed in a little bit from other authors. Of his own there is, on the other hand, nothing" (Wolff 1764, 102).

Wolff long held out hope that Haller would see the light in the experimental findings and return to an epigenetic orientation. Only upon the publication of Volume 8 of Haller's *Elementa*, where the issues of generation were treated, did Wolff see how intransigent Haller had become. "There is nothing of epigenesis," the title of Haller's chapter read. He went on to propound the theological origin of all generation in a balder form than anywhere else in his works, adding a ridiculous calculation of all the humans that must have been encapsulated at the original creation of Eve (Haller 1766). Bonnet, to be sure, might have relished this text. Wolff simply gave up eliciting Haller's approval. In particular, Wolff's response to Haller's irritated warning in 1766 that epigenesis was dangerous to religion suggests that by that point he came to recognize an ideological dimension that went beyond the experimental uncertainties and theoretical divergences that had driven their controversy earlier (see Schuster 1941).

The last thing Wolff needed, in his professional difficulties, was to be lumped with materialists and atheists. He answered the Haller–Bonnet religious scruple in a manner that was strikingly in line, nonetheless, with Maupertuis. A rational-scientific explanation of the world, in his view (as in that of Maupertuis), in no way gainsaid the need for an intelligent creator at its origin. While he admitted that preformation was peculiarly suited to religious exploitation, he denied that epigenesis threatened religion. As he wrote in his letter to Haller, "against the existence of divine Power, nothing is demonstrated if bodies are produced by natural forces and natural causes; for these forces and causes themselves and indeed nature itself, require an author in the same way as organic bodies" (Wolff to Haller, cited in Schuster 1941).

Wolff suspected Bonnet had an influence on the hardening of Haller's position – a suspicion that we can now verify from the Bonnet–Haller correspondence and from Bonnet's extensive invocations of Haller in his works of the 1760s and even more in their revised editions published after Haller's death (Sonntag ed., 1983; Bonnet 1776; 1782). Bonnet was a preformationist true believer and by 1760 he did not really *care* what Wolff had found, since the whole matter had been settled by Haller's work on the chicken embryo, and nothing Wolff could do as an experimental embryologist could match the incomparable Haller: case closed. During his last years in Berlin Wolff worked exigently on the formation of the intestine in the chicken embryo to prove he was right.

Wolff decided to accept a call to the St. Petersburg Academy of Sciences in 1766–1767 (Roe 1979, 4–5). He had been recommended to the Russian authorities as early as 1760 by Leonhard Euler, who clearly knew of and praised highly his dissertation work (Jahn 1998/1999, 43n). Euler was one of Maupertuis's closest allies, assuming direction of the Prussian Academy upon Maupertuis's departure in 1756 and especially after his death in 1759. Wolff only published his work on the formation of the intestine in St. Petersburg. Haller at least bothered to read it. Bonnet never did (though they were in Latin); it was enough that Haller dismissed them. Karl Ernst von Baer, certainly an experimental biologist of a class with Haller, had the last word (Baer, 1828), but it was too late for Wolff ever to hear it.

Johann Nicolaus Tetens

The *Philosophische Versuche über die Menschliche Natur und Ihre Entwickelung* (1777) of Johann Nicolaus Tetens showed a remarkable attunement to the essential issues of his time across a wide variety of fields. In particular, Tetens became the most discriminating German philosophical commentator on the controversy between Bonnet and Wolff over preformation and epigenesis. Tetens sought to find some middle ground between the extreme positions identified with Bonnet (preformation, or, as Tetens and his age termed it, *evolution*) and Wolff (epigenesis) (Tetens 1777, II, 460–500). He took a very positive view of Bonnet's position, only faulting it, quite typically, for going too far. He believed that Bonnet had made a compelling case for the necessity of a preexisting *form* for the reproduction of organisms; otherwise, the species continuity and the internal coherence of each new specimen seemed to him too inexplicable. On the other hand, he believed epigenesis proposed a sound claim that organisms developed, incorporating into themselves alien, even inorganic materials, and creating new forms, even developing through stages, especially in the embryo, which would be displaced in the ultimate, mature organism. The question of "new forms" was, he asserted, the essentially contested issue between *evolution* and *epigenesis*. An adequate theory would have to incorporate both the need for a guiding design and the fact of emergent properties.

Significantly, Tetens found Wolff's experimental results unconvincing, and turned instead to the earlier ideas of Buffon as a more adequate representation of the merits of epigenesis. What he liked about Buffon's version of epigenesis was the latter's insistence on a "interior mold," a preexisting, necessary form for the organization of the "organic molecules" of each new organism. Thus Tetens proposed a synthesis: "evolution via epigenesis" – "an evolution from within which can occur through epigenesis" (Tetens 1777, II, 498). Thus, "this epigenesis through evolution appears to be the general form of emergence of organized beings" (Tetens 1777, II, 500). Based on a fixed original form, the organic matter of a new organism formed itself and grew into the mature specimen via the mechanical processes characterized by epigenesis. This reformulation not only incorporated the sophisticated revision of preformation initiated by Haller and Bonnet, but also seems clearly to anticipate Kant's notion of "generic preformation."

Johann Gottfried Herder

In his decisive discussion of epigenesis in *Ideen* (1784), Herder referred explicitly to the work in generation theory and embryology of William Harvey and Caspar Friedrich Wolff:

> How must the man have been astonished, who first saw the wonders of the creation of a living being! Globules, with fluids shooting between them, become a living point; and from this point an animal forms itself. The heart soon becomes visible, and, weak and imperfect as it is, begins to beat.... What would he who saw this wonder for the first time call it? There, he would say, is a living organic power: I know not whence it came, or what it intrinsically is, but that it is there, that it lives, that it has acquired itself organic parts out of the chaos of homogeneous matter, I see; this is incontestible.
>
> If we contemplate these changes, these living operations, as well in the egg of the bird as in the womb of the viviparous quadruped, then, it seems to me, one is not being forthright [*spricht man uneigentlich*] if one talks of germs [*Keimen*] that are only developed or of an *epigenesis* according to which the members would accrete externally [*die Glieder von aussen zuwüchsen*]. It is a matter of *formation* [*Bildung*] (genesis), an effect of internal forces for which Nature has prepared the raw materials [*Masse*] which they incorporate into themselves in order to make themselves visible.
>
> (Herder 1784, 273–274, 173)

That is, there existed an "internal nature [which] becomes visible in a mass appertaining to it, and must have the prototype of its appearance in itself, whence or wherever it may be" (Herder 1784, 274). Herder denied individual preformation (*das Evolutionssystem – emboitement*) and merely external, mechanical change (reductive mechanism) as sufficient explanations, and argued that there had to be an immanent force behind such variation. He called it "genetic force."

Thus, Herder elaborated into a general interpretive principle the embryological idea of epigenesis he took from Wolff.[5] He asserted the emergence of increasing complexity and differentiation as an immanent principle of natural development, as an intrinsically *historical* character/tendency of the entire physical world. "Quite generally, nothing in nature is separated, everything flows onto and into everything else through imperceptible transitions" (Herder 1778, 195). "Everything in nature is connected: one state pushes forward and prepares another" (Herder 1784, 194). "Nothing in nature stands still" (Herder 1784, 177). "Inferior powers ascend to the more subtle forms of vitality" (Herder 1784, 178).

Herder articulated this developmental propensity in nature in a theory of the world as composed primarily of "forces" (*Kräfte*) organized hierarchically (Clarke 1942). "The one organic principle of nature … we here term *plastic*, there *impulsive*, here *sensitive*, there *artful* … is at bottom but one and the same organic power" (Herder 1784, 102). He proposed to discern morphological universals: "[I]n marine life, plants, and even inanimate things, as they are called, one and the same groundwork may prevail, though infinitely more rude and confused" (Herder 1784, 66–67). Organic form was continuous with the inorganic. "The active powers of Nature are all living, each in its kind; they must possess a something within, answerable to their effects without, as Leibniz advanced, and as all analogy seems to inform us" (Herder 1784, 98).

> The new discoveries that have been made respecting heat, light, fire, and their various effects on the composition, dissolution, and constituent parts of terrestrial substances, the simpler principles to which the electric matter, and in some measure the magnetic, are reduced, appear to me … at least considerable advances which will in time enable some happy genius … to explain our geogony on principles as simple as those to which Kepler and Newton have reduced the solar system.
>
> (Herder 1784, 22)

Herder sought to explain both the physical and the moral world, both nature and spirit, in terms of attractive and repulsive forces which were unanalyzable but actual, efficient causes (Proß 1987, 851). One found dynamic polarity, Herder wrote,

> spread throughout the whole world order. Everywhere two forces set against one another which nonetheless must work together and in which only by the combined and appropriate influence of both emerges the higher reality of a wise order, development, organization, life. All life arose in such a manner from death, out of the death of lesser forces, all wholes of order and of design from light and shadow, out of diverging, mutually opposing forces.… Mathematics, physics, chemistry, physiology of living beings all seem to me to provide evidence for this everywhere.
>
> (Herder 1775, 536–540)

In 1787 Herder claimed his reworking of Spinozism provided a coherent interpretation which would bring an end to all the objectionable expressions of how God, according to this or that system, may work on and through dead matter:

> It is not dead but lives. For in it and conforming to its outer and inner organs, a thousand living, manifold forces are at work. The more we learn about matter, the more forces we discover in it, so that the empty conception of a dead extension completely disappears. Just in recent times, what numerous and different forces have been discovered in the atmosphere! How many different forces of attraction, union, dissolution and repulsion, has not modern chemistry already found in bodies?
>
> (Herder 1943, 105)

Herder's vitalist materialism thus invoked the most important recent developments in the natural sciences, especially in the fields of electricity, chemistry and physiology. As Elias Palti has noted, "the study of the natural sciences of his time clarifies fundamental aspects of Herder's historical view, and, conversely, the analysis of Herder's philosophy allows us to better understand [the developments in natural science]" (Palti 1999, 323n).

Herder was suggesting that the proper sense of epigenesis was a fertile and unpredictable creativity in nature, a sweeping notion of its pervasive, fundamental "genetic force." He was not rejecting but rather radicalizing the idea of epigenesis. Kant understood exactly:

> As the reviewer understands it, the sense in which the author uses this expression [i.e., *genetische Kraft*] is as follows. He wishes to reject the system of evolution on the one hand, but also the purely mechanical influence of external causes on the other, as worthless explanations. He assumes that the cause of such differences is the vital principle [*Lebensprinzip*] which modifies *itself* from within in accordance with variations in external circumstances, and in a manner appropriate to these.
>
> (Kant 1784/1785, 62–63)

This was far too open-ended for Kant, and he insisted that this "genetic force" was not unlimited, and could not lead to a mutation of species. A sensible interpretation, Kant added, should simply accept this capacity for variation as "germs" (*Keime*) or "original endowments" (*natürliche Anlagen*) incapable of further determinate elucidation (Kant 1784/1785, 63).[6] Otherwise, the very notion of species distinction was at risk. To be sure, Herder was careful to repudiate such a speculation explicitly: "In truth, ape and man were never one and the same species" (Herder 1784, 229–231). But that could not disguise for Kant the radical potential latent in Herder's text; indeed, this was its historical impact for "attentive and adept readers," as Heinz Stolpe has acutely observed (Stolpe 1964, 315). Charlotte von Stein wrote to this effect in 1784 in a letter to Goethe's friend Knebel: "Herder's text makes it appear that we started out as plants and animals" (cited in Wenzel 1990, 137). H. B. Nisbet, too, emphasizes that it was Kant's recognition of the potential in Herder's *type* theory for a *transmutation* theory that led to his critique (Nisbet 1970, 109). Herder deliberately set about erasing the border lines Kant had so carefully drawn not only between life and matter but, even more grievously for Kant, between animal and man (Zammito 1998, 2003; Sloan 2002). This was Herder's whole point: in Goethe's words, "nothing [physically] specific could be found to differentiate between man and animals" (Goethe 1784, cited in Irmscher 1987, 70). This was Herder's salience for late eighteenth-century German science: no one articulated with the same breadth and vivacity as Herder the prospect of confirming that continuity.

Immanuel Kant

In his *One Possible Basis for a Demonstration of the Existence of God* (1763), Kant had already addressed the new twist toward epigenesis introduced by Maupertuis and Buffon, declaring it far-fetched and doomed, since it ascribed far too much power to mere matter, or what Kant dismissed as "hylozoism":[7]

> The internal forms proposed by *Buffon*, and the elements of organic matter which, in the opinion of *Maupertuis*, join together as their memories dictate and in accordance with the laws of desire and aversion, are either as incomprehensible as the thing itself, or they are entirely arbitrary inventions [*sind entweder eben so unverständlich als die Sache selbst, oder ganz willkürlich erdacht*].
>
> (Kant 1763, 115)

Epigenesis as an empirical scientific theory had *no* prospect of realization for Kant, because he held firm to the conviction that "one is incapable of rendering distinct the natural causes which bring the humblest plant into existence" (Kant 1763, 138). Thus, for Kant, the hypotheses of Buffon and of Maupertuis were *not* scientific but only fanciful or metaphysical, i.e., *ganz willkürlich erdacht*. He allowed no prospect, notwithstanding the purported superiority of the scientific *motivation* of their enterprise, of any evidentially warranted *explanation*.

Kant unequivocally affirmed living forms as prime "examples of an artificially devised order of nature" (Kant 1763, 136n). Thus, life had to be recognized as "a contingent, purposeful phenomenon," i.e., one that was the direct "product of choice" (Kant 1763, 96):

> The structure of plants and animals … cannot be explained by appeal to the universal and necessary laws of nature.… [I]t would be absurd to regard the initial generation of a plant or animal as a mechanical effect incidentally arising from the universal laws of nature.
>
> (Kant 1763, 114)

For example, "no general cause can be adduced to explain … the retractability of the claws of the cat" (Kant 1763, 114). As for unity of the order of nature, Kant could only rationalize that "the creatures of the plant- and animal-kingdoms everywhere offer the most admirable examples of a unity which is at once contingent and yet in harmony with great wisdom" (Kant 1763, 107).

Many scholars sense a shift in Kant's position on epigenesis as a theory of embryology by the critical period, but the effort to account for it has not occasioned unanimity. The key may well have been found in Kant's reception of Johann Nicolaus Tetens. In an exemplary philological consideration of the sources Kant appeared to have drawn upon to formulate his own conceptions, Jennifer Mensch has shown that the translations of Buffon and Bonnet in the 1750 and 1760s respectively blurred the French originals as between two crucial German terms – *auswickeln* and *entwickeln*. "Buffon's German translators used *auswickeln* for *développer* during Buffon's discussion of generation, whereas Bonnet's translators tended to use *entwickeln* for *développer*" (Mensch 2014, 201, n238). She continues:

> Kant's own patterns of word choice reveals, to some extent, his linguistic influences. Up until 1777 Kant used *Auswickelung* some 14 times, compared to 11 uses of *Entwickelung*. After 1777 Kant used *Auswickelung* 9 times, but *Entwickelung* would appear some 112 times.
>
> (Mensch 2013, 201, n238)

She infers that in his early race essay, Kant was reading Buffon far more than Bonnet. But everything shifted after his intense study of Tetens. It would appear that it was Tetens who helped Kant to a determination of his stance not only on the specific debate between Haller/Bonnet and Wolff, but on the general question of preformation and epigenesis as well, namely a *fusion* of the two positions. *Philosophische Versuche* (1777) had immediate and widespread impact on the German philosophical community, and especially on Kant. J. G. Hamann reported in a letter to his friend Herder (May 17, 1779) that in the years leading to the completion of his *Critique of Pure Reason* (1781) Kant had this work by Tetens constantly open on his writing desk. The impact on this particular issue appears quite clear. Kant self-consciously elected to adopt the notion of *Entwickelung*, with its strongly epigenetic connotation. But – like Tetens, and perhaps persuaded by him – he was convinced that this development was always constrained by a prior formative order, hence his 1790 formulation of "generic *preformation*" as the only viable meaning for epigenesis as a theory of generation (Kant 1790, 423). With this strong sense that epigenesis could not be unconstrained, Kant was predisposed to find Herder's invocation of epigenesis in the *Ideen* "reckless" and in need of strict reproof. But that reproof, I suggest, might well be extended to the *source* of Herder's notions, as well, namely Caspar Friedrich Wolff.

There is a striking absence of explicit reference to Wolff in Kant's entire oeuvre (Goy 2014, 44n). That certainly does not mean that Kant was unaware of Wolff, but it does raise questions about the centrality of Wolff's ideas in Kant's understanding of epigenesis. This is especially pertinent because a number of recent interpreters have contended that Wolff's views of epigenesis were decisive in enabling Kant to adopt this view of animal generation (Dupont 2007; Huneman 2007b). With Ina Goy, I find these reconstructions implausible (Goy

2014). Even more strongly than Goy, I take Wolff instead to represent a direction in the empirical theory of animal generation and epigenesis that was far more mechanistic and materialistic than Kant could support. Indeed, I believe that it was Herder's adoption of these impulses in Wolff that alarmed Kant in his reviews of the *Ideen*. Hence, it is crucial to establish what Kant did know of Wolff, and how he incorporated it in his thought.

Jean-Claude Dupont adopts the view that it was Wolff's 1768/1789 text, *De Formatione Intestinorum*, that "made realization of the Kantian teleomechanistic research program an actual possibility" (Dupont 2007, 37). Historically, I think this is wrong. I do not believe that Kant had any direct or even *derivative* awareness of this text or its salience as, indeed, "the first great text of modern embryology" (Dupont 2007, 37). Quite simply, Dupont has no textual basis for his claim that Wolff "was a major reference in this new epistemological situation that Kant had to face" (Dupont 2007, 47). Ina Goy has proven that. Thus, it is historically wrong to claim that, "after Wolff, empirical investigations into the mechanisms of this new 'generic preformation' (Kant) would become possible" (Dupont 2007, 47). There is simply no evidence that the latter was the source of any significant shift in Kant's position on epigenesis.

Moreover, Dupont shares with Philippe Huneman a conception of Wolff's *vis essentialis* and of epigenetic embryology that I think goes further in the direction of vitalism than is justified by the texts. Dupont is more equivocal on this matter, yet he claims, "Wolff remained very close to Blumenbach" (Dupont 2007, 47). This strangely inverts their historical precedence, but, more gravely, I believe, it misunderstands the aversion Blumenbach felt toward Wolff's theory, and that Kant, without mention of Wolff, shared, namely against the *hylozoistic* potential of mere matter to generate life.[8] Dupont comes close to this point:

> the *vis essentialis* was driven by the qualities or the properties of the matter that are passed on from generation to generation.... While Wolff accepted the idea of mechanical forces or powers as the source of life processes in general, and specific to them, there was no idea of an immanent teleological power.

For Wolff, *vis essentialis* was a kind of mechanical force with effects "of the attraction and aversion type" (Dupont 2007, 47). That is a long way from Blumenbach, and *a fortiori* from Kant.

Huneman avows that Wolff's notion of epigenesis "enabled Kant to resolve the philosophical problem of natural generation ... to determine what is proper to the explanation of living processes" (Huneman 2007a, 73). He takes Kant to have formulated what he calls a "generation dilemma" in the *Beweisgrund* text of 1763, namely that neither preformation nor epigenesis as then formulated could offer satisfactory accounts of animal generation (Huneman 2007a, 84). That is entirely correct. But Huneman then argues that it was the empirical embryology of Wolff that offered Kant a path out of this impasse. That I find implausible, for all the reasons articulated above. Moreover, I find Huneman's reconstruction of Wolff's theory quite problematic. According to Huneman, "essential force … has special status. It is not a natural, Newtonian force, as was Haller's *vis insita*.… It is not the cause of some regular effects accorded by a mathematical law" (Huneman 2007a, 83–84). This was because embryology entailed discontinuities, which obviated direct causal relations and required instead an "epistemological" guarantee to secure the identity of the organism across these discontinuities. This, Huneman argues, was the role of *vis essentialis* in Wolff. He argues: " 'causes' must not be taken in a Newtonian sense, as referring to a mathematical law; rather, it should be read historically, as a retrospective reconstruction of a series' order" (Huneman 2007a, 84).

There is something right and something wrong in all this. First, to take "cause" as requiring a mathematical law is to restrict unduly the Newtonian notion, leaving out the crucial elaboration of the *Queries* to the *Optics*, which became the basis for experimental Newtonianism in the eighteenth century. And, second, *historical* causality, which was indeed crucial for the constitution of both geology and life science over the course of the eighteenth century, was *not* an idea that fit at all well into Kant's philosophy of science (Kant 1786, 467–469). Even had Kant discerned this in Wolff, which I dispute, he would not have affirmed it. Thus, I reject Huneman's conclusion that "although Kant is inspired by Blumenbach's doctrine, his theory draws on the new epistemological territory (which yields this doctrine as well as its Wolffian alternatives) settled by Wolff" (Huneman 2007a, 88). Rather, I stress what Huneman himself concedes: "Kant in his own biological theories was not as epigenetic as Herder or even Caspar Wolff" (Huneman 2007a, 89).

Goy persuasively argues that Wolff's "*vis essentialis* accounts for mechanic [*sic*] effects in matter, whereas, according to Kant, the formative power explains the intentional order (form, end, purpose) of an individual organized being, its parts and its species" and "this view is closer to Blumenbach than to Wolff" (Goy 2014, 44). But she goes on to note that Wolff articulated a theory of the organism in terms of parts and whole which resonates with Kant's classic formulation in the *Critique of Judgment* and which was never a matter that Blumenbach elaborated (Goy 2014, 44). Thus, she finds a connection to Wolff in this regard. But she herself makes two crucial points that undermine that link. First, there is no textual evidence that Kant noted this element in Wolff's thought (Goy 2014, 44n). Second, Wolff argues precisely the opposite position with regard to parts and whole in the organized being from that which was central to Kant's theory. Wolff's claim was that the whole was only sequentially constructed from the parts, while Kant insisted upon the mutual constitution of parts and whole (Goy 2014, 58). Wolff's view represents exactly what Peter McLaughlin has reconstructed as the decisive meaning of mechanism in Kant's *Critique of Judgment*. Mechanism, however legitimately scientific in its maxim, could not overcome the limits of human understanding and therefore needed teleological supplement (McLaughlin 1990). Such ideas would have been utterly unacceptable to Wolff, just as, I contend, Wolff's mechanistic model of parts and whole could not have been of any use to Kant.

Blumenbach between Kant and Wolff

How did Blumenbach respond to Kant's appropriation of his ideas in the essay "On the Use of Teleological Principles in Philosophy" (1788)? Blumenbach's first major publication after Kant's essay appeared, the third edition of the *Handbuch der Naturgeschichte*, was dated March 1788, and it unsurprisingly gives no evidence of Blumenbach's attention to Kantian ideas (Blumenbach 1788). But, less than a year later, in January 1789, he published his revised version of *Über den Bildungstrieb* and sent Kant a copy of this work in acknowledgment of Kant's references to him in the 1788 essay (Blumenbach 1789).[9] The preface to this second edition of his essay on the *Bildungstrieb* advised readers that his earlier version was "immature" (Blumenbach 1789, A4).

What did Blumenbach intend by his preface of January 1789, and by routine appeals in later versions of his *Handbuch* and of his dissertation on human variety, "not to confuse [this second edition] with the immature treatise that appeared under a similar title in 1781" (Blumenbach 1789, A4)?[10] Can we take it for granted that this was "immaturity" by Kantian standards? Timothy Lenoir explicitly claims that Blumenbach's "mature formulation resulted from his encounter with Kant's work" (Lenoir 1980, 84n).[11] That is not historically defensible for the preface of January 1789, and it is quite problematic for later incorporations of Kantian language. I suggest that we must regard Blumenbach's judgment of his earlier work in a more complex light. He was already making changes in his 1788 *Handbuch*, before we have any reason to suspect Kantian influence. He had encountered significant resistance to his ideas – and from two fronts: the die-hard preformationists (Bonnet, Spallanzani, Caldani), but also the more aggressively naturalistic epigenesists – Thomas Sömmerring and Georg Forster and, of far greater importance, Caspar Friedrich Wolff.[12] If we consider the texts of 1781 and 1789 in juxtaposition, what is foremost is the clarity with which Blumenbach characterizes his central innovation. The structure of the argument is considerably clearer: after the historical background leading up to his own discovery, Blumenbach presents a thorough drubbing of the arguments for preformation, followed by a clear account of the advantages of his *Bildungstrieb* theory. He is far more comfortable that he has made a major breakthrough and that he has defeated his rivals on that front. That is, Blumenbach believed he had dramatically improved the exposition of his *scientific* position by 1789, not – or not *just* – his sophistication about philosophy of science.

One of the most important aspects of his argument in 1781 was that the *Bildungstrieb* encompassed and explained three vital functions – generation, nutrition and regeneration. In the 1789 version, nutrition gets scant attention. It is generation and regeneration that Blumenbach believes offer the best support for his theory in comparison with others. But it may also be that he had addressed the nutrition issue separately, in a prizewinning essay submitted to a competition sponsored by the Academy of Sciences in St. Petersburg in 1782, and presided over by his rival epigenesist, Caspar Friedrich Wolff (Wolff 1789b). While Wolff awarded Blumenbach the prize, he published a far lengthier work of his own on the topic, taking a sharply critical posture toward Blumenbach's views (Wolff 1789a). There are thus grounds to think that there was another presence besides Kant whose appraisal of his work loomed large for Blumenbach in the later 1780s, namely Wolff.

And this might well explain the most important methodological clarification in the 1789 version, which did bring Blumenbach happily into alignment with Kant: the radical separation of organic from inorganic form and the repudiation of any hylozoism. Blumenbach embraced a fundamental ontological distinction between the general order of nature and the specific order of the organic. In the 1789 version of his *Bildungstrieb* book, Blumenbach made this very clear: "No one could be more totally convinced by something than I am of the mighty abyss which nature has entrenched [*befestigt*] between the living and the lifeless creation, between the organized and the unorganized creatures" (Blumenbach 1789, 79). Indeed, Blumenbach shared Kant's skepticism about a bridge from the inorganic to the organic and about the phylogenetic continuity of life forms. What bound them most together was their commitment to the fixity of species and their rejection of the reality of the *scala naturae*. Yet Blumenbach drew neither of these commitments from Kant. They were already expressed with clarity in his dissertation of 1775 and especially the first edition of his *Handbuch* of 1779. These were basic issues for anyone taking up natural history or life science in the eighteenth century. It is far more likely that Blumenbach adopted them from Albrecht von Haller than from Kant.[13] What remains is to consider whether the *reasons* for Blumenbach's commitments were the same as the reasons for Kant's commitments to these same positions.

When he first presented his notion of the *Bildungstrieb*, Blumenbach concentrated on how it answered certain physiological problems in organisms better than the alternative theories of preformation and of epigenesis. He did not dwell yet on the methodological or epistemological status of his concept. In the 1782 edition of his *Handbuch der Naturgeschichte*, his treatment of the idea of the *Bildungstrieb* once again gave no attention to this epistemological issue. He simply carried forward with his empirical exposition. Perhaps he came to regard this as one of the "immature" features of his work. He changed already in the 1788 edition of the *Handbuch* – presumably before he could have absorbed very much of Kant's methodological thinking. He introduced a new section, immediately after defining his *Bildungstrieb*, with the following language:

> The *cause* of this formative drive can admittedly be so little adduced as that of attraction or gravity and other such generally recognized natural forces. It is enough that it is a distinctive force whose undeniable existence and broad influence throughout all of nature is revealed by experience, and whose constant phenomena offer a far more ready and clear insight into generation and many other of the most important topics of natural history than other theories offered for their explanation.
>
> (Blumenbach 1788, 14)

There is, here, a tacit Newtonian analogy, without the mention of Newton by name. Moreover, the argument is presented in terms of the *general* order of nature: no strong distinction is made between the organic and the inorganic realms in terms of the nature of such forces, though, clearly, this particular force operates in generation and organic forms.

In the second edition of his *Bildungstrieb* book, Blumenbach became far more explicit about the Newtonian connection:

> The *term* formative drive, just like the *terms* attraction and gravity, etc. serve no more and no less than to denote a force whose constant effect is recognized but whose *cause* just as little as the causes of the other, nonetheless so generally recognized natural forces, remains for us a *qualitas occulta*. That does not hinder us in any way whatsoever, however, from attempting to investigate the effects of this force through empirical observations and to bring them under general laws.
>
> (Blumenbach 1789, 32–33)

In the attached footnotes, Blumenbach referred directly to Newton, and then, in the context of the phrase *qualitas occulta*, to Voltaire's exposition of Newton, in particular to the passage where Voltaire argued that from a mere "blade of grass" to the order of the stars, *all* causes (physical as well as biological) were simply occult qualities (Blumenbach 1789, 32n, 33n).[14] This was standard epistemology of science in the wake of John Locke's discrimination of "nominal" from "real" essences, of empirical (external) observation from the "inner" or ultimate reality of nature (Locke 1689). It is important to stress that Kant was hardly a necessary influence for Blumenbach in making this Newtonian appeal. It was common practice among all innovative life scientists. Haller and Buffon had done it, and so had Caspar Friedrich Wolff (Wolff 1764; see, e.g., Roe 1979; Gaissinovich 1968). As Peter McLaughlin has argued, making the Newtonian appeal was constitutive for the emergent life sciences in the late eighteenth century (McLaughlin 1982). While it was epistemologically expedient, this may well have been disingenuous in many cases, for the forces were taken quite straightforwardly as real, even if the *ultimate* causes remained mysterious. That anything like Kant's critical epistemology was in play must be open to considerable doubt.

Ultimately, then, what major break was there between Blumenbach's 1781 formulation and the new "mature" formulation of 1789? McLaughlin has set this inquiry on the proper path by a very close reading of Blumenbach's various formulations of the notion of the *Bildungstrieb* in successive publications. As McLaughlin is quite right to maintain, Blumenbach did not do a very good job in explicating his *Bildungstrieb*: "what that is supposed to mean exactly is nowhere systematically elaborated" (McLaughlin 1982, 364).[15] For McLaughlin, the contrast with C. F. Wolff is most illuminating, and the issue of the relation of the formative drive to organic matter is central. I think that is exactly the right line of attack, though I deviate somewhat from McLaughlin in the interpretation of these matters.

In his *Handbuch* of 1782, Blumenbach wrote that "a particular, innate *drive*, active throughout its life, lies in every organized body" (Blumenbach 1782, 15). In the *Handbuch* of 1788, he wrote that one could find "throughout all nature the most unmistakable traces of a virtually general drive to give matter a determinate form, which already in the inorganic realm is of striking effectiveness" (Blumenbach 1788, 12–13). As McLaughlin properly observes, "In fact, the *only clear substantive difference* in the key formulations of the theory of the *Bildungstrieb* between the 'more mature' and the 'immature' phase is the replacement of an 'innate' drive by a 'general' drive" (McLaughlin 1982, 371). As the editor of the reprint of Blumenbach's classic comments, though Blumenbach called the earlier version immature, "nevertheless even stylistically the essential statements are hardly changed" in the later ones (Karolyi 1971, vi). What Karolyi discerns is a clearer self-assertion versus Haller and Wolff, but "argumentation, examples, and the core of the statement remain unchanged" (Karolyi 1971, xii). There were, to be sure, "in part more refined, more differentiated formulations and some additions to the exposition of the first edition," but for Karolyi these hardly amounted to the "completely new construction of the theme" alleged by Robert Herrlinger in his preface to the reprint of the work of C. F. Wolff (Karolyi 1971, xi). Herrlinger had implied that Blumenbach *needed* such a new formulation in light of Wolff's criticisms (Herrlinger 1966, 19n). But Herrlinger was not entirely off the mark. In short, the tension between Blumenbach and Wolff, more than his affinity to Kant, occasioned Blumenbach's discomfort with the "immaturity" of his work of 1781.

This would make no sense if Blumenbach really believed that C. F. Wolff was a "mystical" vitalist, as Lenoir strangely conceives him to be.[16] Rather, it is the notion of a continuity from the inorganic (in Wolff, the chemical) to the organic – i.e., a materialist naturalism or "hylozoism" – in Wolff that Blumenbach wishes to distance himself from. Blumenbach found Wolff's notion of epigenesis problematic as much – or more – for the metaphysical quandaries as for the methodological ones. McLaughlin identifies crucial changes that Blumenbach introduced in 1791, after he had absorbed Kant's ideas not only from the 1788 essay but from the *Critique of Judgment* that Kant had sent him. As we have noted, in 1788 Blumenbach found that "throughout all nature the most unmistakable traces of a virtually general drive to give matter a determinate form, which already in the inorganic realm is of striking effectiveness" (Blumenbach 1788, 12–13). In 1791, Blumenbach pruned the line as follows: one finds "in the entirety of organic nature the most unmistakable traces of a generally distributed drive to give matter a determinate form" (Blumenbach 1791, 14). The appended clause from 1788 was eliminated altogether. In 1789, as we have noted, Blumenbach compared the *Bildungstrieb* to "the *terms* attraction and gravity … generally recognized natural forces." But in 1797 he changed this to: "The term *Bildungstrieb* just like all other life forces" (Blumenbach 1797, 18). The point, here, is that Blumenbach wished his formative drive to be considered only in comparison with other *life forces*. The thrust, as McLaughlin notes, was to make a radical distinction between the organic and the inorganic realms and to assign the drive exclusively to the former.

McLaughlin derives from this shift in position in Blumenbach by 1791 that the *Bildungstrieb* is not the *cause* of life but rather its *consequence* (McLaughlin 1982, 359). That is, what all the earlier (materialist/naturalist) proponents of epigenesis sought to explain (life as an emergent property arising out of matter itself) in Blumenbach becomes an inexplicable presupposition. For La Mettrie, Buffon and Holbach, according to McLaughlin, "life was a mechanical result of organization" – that is, of the general order of nature grounded in physics and chemistry. Blumenbach, by contrast, aimed "to explain organic *form* through organic *matter*." That is, an *organic* force is "a force that only *has effect within* organic matter, not a force that somehow *causes* the transition from inorganic to organic matter." The *Bildungstrieb* did not explain life but rather presumed it (McLaughlin 1982, 357).[17] While there was organization already in inorganic matter, there was something extra about organic matter, which John Hunter called a "supplementary force," something "applied in addition" (McLaughlin 1982, 359).[18]

By contrast, for McLaughlin, "Wolff's essential force was a *chemical* attraction-repulsion force" (McLaughlin 1982, 365). Thus, for Wolff, matter was heterogeneous, i.e., it achieved various *levels* of organization, and, once it passed a certain threshold, there ensued something of a chemical chain reaction that initiated life. The important inquiry was into the component constraints that directed the chain reaction. For Blumenbach, by contrast, McLaughlin believes the important question was the *inherent* relation between a distinctively organic matter and the forces unique to it. That did not mean one could not draw *analogies* from the inorganic to the organic, for, Blumenbach wrote,

> even in the inorganic realm the traces of formative forces are so unmistakable and so general. Of formative *forces* – but not by far of the formative *drive* (*nisus formativus*) in the sense this term assumes in the current study, for it is a *life* force [*Lebenskraft*] and accordingly as such inconceivable in inorganic creation – rather of other formative forces, which provide the clearest proof in this inorganic realm of nature of determinate and everywhere regular formations [*Gestaltungen*] shaped out of a previously formless matter.
>
> (Blumenbach 1789, 80; my emphases)

This distinction between the formative *forces* [*Kräfte*] that structure the inorganic realm and the formative *drive* (*Trieb*; note that it is always singular in Blumenbach's usage) that is unique to organic life, and indeed a *Lebenskraft* among others, proved crucial for Kant.

This was what Kant found most gratifying in the new book, as he reported in his letter of acknowledgment to Blumenbach, August 5, 1790 (Kant, B, AA:11, 176–177). In the *Critique of Judgment* he elaborated:

> Blumenbach … rightly declares it to be contrary to reason that raw matter should originally have formed itself in accordance with mechanical laws, that life should have arisen from the nature of the lifeless, and that matter

> should have been able to assemble itself into the form of a self-preserving purposiveness by itself; at the same time, however, he leaves natural mechanism an indeterminable but at the same time also unmistakable role under this inscrutable *principle* of an original *organization*, on account of which he calls the faculty in the matter in an organized body (in distinction from the merely mechanical *formative power* [*Bildungskraft*] that is present in all matter) a *formative drive* [*Bildungstrieb*] (standing, as it were, under the guidance and direction of that former principle).
>
> (Kant 1790, 424)

This passage in the *Critique of Judgment* makes the distinction between formative force and formative drive prominent.[19]

In my view, Kant is simply appropriating Blumenbach for philosophical purposes alien to Blumenbach's own scientific practice. Blumenbach *never* considered his formative drive anything but an actual force in nature. To be sure, he found Kant's suggestion that he brought teleological and mechanical explanations together in his scientific practice quite pleasing, but it is not clear that he understood Kant's painstaking argument for their radically different roles in scientific explanation. In short, notwithstanding Lenoir (and Jardine), Blumenbach's affiliation with Kant is best understood as a *misunderstanding*. Moreover, it was a *creative* misunderstanding, because it enabled Blumenbach and his followers to continue with even greater energy the development of that new science of *Naturgeschichte*, that "daring adventure of reason," that Kant by 1790 found deeply problematic.

Conclusion

Caspar Friedrich Wolff presented eighteenth-century German philosophy with the most dramatic challenge to entrenched ideas about the continuity of inorganic and organic matter ("hylozoism"), and therewith raised fundamental issues about materialism, vital force and the specific character of organisms and their generation. In no small measure thanks to Wolff, epigenesis proved the cardinal challenge to philosophical considerations of life in the eighteenth century and thus the indispensable trigger for the theoretical elaborations of what would become the science of biology in the nineteenth century.

Notes

1 Roe (1979, 1) distinguishes between opposition to preformationism and advocacy of epigenesis: "Buffon, Maupertuis, Needham, and the materialists La Mettrie and Diderot" all opposed preformationism; "not all of these can be classified as true epigenesists, however" (ibid., 3n). That, of course, depends on how one characterizes "true epigenesis."

2 As Roe notes, "Wolff's dissertation was unusual in length, originality of content, and the fact that the name of his dissertation director was not printed on the title page…" (Roe 1979, 5n).

3 Haller 1760. The opening line of his review is widely cited, and it was welcomed with great joy by Wolff. What followed is a different matter. See Schuster 1941.
4 The first letter to Haller, which accompanied a copy of the dissertation, gives clear indication of Wolff's hopes. The dissertation itself has no armature of self-defense against so formidable an opponent, which prudence would have dictated, had he anticipated such opposition. The later *Theorie von der Generation* (1764) demonstrates that Wolff was quite capable of developing such a defense – indeed, a counterattack – when he understood its necessity.
5 H. D. Irmscher (1973, 17–57) notes Herder's early and distinctive embrace of the idea of epigenesis. Palti suggests a more ambivalent relationship, offering a number of distinctions and tensions in the biological theories and in Herder's reception of them, which he conceives as "uneven developments" (Palti 1999, *passim*). For more on this, see Zammito 2001.
6 Kant wrote:

> The reviewer is fully in agreement with him here, but with this reservation: if the cause which organises *from within* were limited by its nature to only a certain number and degree of differences in the development of the creature which it organises (so that, once these differences were exhausted, it would no longer be free to work from another archetype [*Typus*] under altered circumstances), one could well describe this natural development of formative nature in terms of germs [*Keime*] or original dispositions [*Anlagen*], without thereby regarding the differences in question as originally implanted and only occasionally activated mechanisms or buds [*Knospen*] (as in the system of evolution); on the contrary, such differences should be regarded simply as limitations imposed on a self-determining power, limitations which are inexplicable as the power itself is incapable of being explained or rendered comprehensible.
>
> (Kant 1784/1785, 63)

7 Maupertuis was the early Kant's paradigmatic instance of a modern hylozoist. Kant's *Dreams of a Spirit-Seer* treats him in exactly that context:

> *Hylozoism* invests everything with life, while *materialism*, when carefully considered, deprives everything of life. Maupertuis ascribes the lowest degree of life to the organic particles of nourishment consumed by animals; other philosophers regard such particles as nothing but dead masses, merely serving to magnify the power of the levers of animal machines.
>
> (Kant 1766, AA:2, 330. See Zammito 2006)

8 Blumenbach wrote: "no one could be more totally convinced by something than I am of the mighty abyss which nature has fixed [*befestigt*] between the living and the lifeless creation, between the organized and the unorganized creatures" (1789, 79).
9 Blumenbach's transmission to Kant in 1789 is acknowledged by Kant in his letter to Blumenbach, August 5, 1790, in Kant 1922, AA:11, 176–177.
10 For later avowals along the same lines see, for instance, Blumenbach 1791, 13; 1797, 17n.
11 Lenoir argues that Blumenbach's "mature theory" was composed only "after he had begun to wrestle with Kant's philosophy of organic form," and ostensibly upon that basis (Lenoir 1980, 83).
12 For one seminal discussion of the epigenesis controversy in Germany, see Roe 1981.
13 Blumenbach commented:

> I think it says a lot – but, as I see it, not too much – when I maintain that Haller was the greatest among all recently deceased scholars who have been working in Europe since Leibniz's death. He was the greatest scholar as concerns variety as well as quantity and depth of his knowledge.
>
> (Blumenbach 1785, 177)

14 One wonders whether it was not this passage from Voltaire that provoked in Kant the famous passage that there would never be a Newton of the blade of grass.

15 Jardine moves too quickly from the correct observation that Blumenbach "offers no positive account of the nature of the formative drive" to the inference that "it is proposed as a heuristic in the search for empirical laws" (Jardine 2000, 26). The Newtonian analogy did not minimize at all the actuality of the formative drive, but only denied access to its *ultimate* cause. This is a vital discrimination if we are to understand the relation between Blumenbach and Kant.

16 I think in several of his publications Lenoir misunderstands Wolff in a manner that sets up his misconstrual of the whole epoch of life science from the late eighteenth to the mid-nineteenth centuries, because he identifies vitalism with "idealism" – i.e., animism. We must rescue "vital materialism" from Lenoir's residual positivism. See Reill 2005. Further, Lenoir imputes to the imaginative construction of hypotheses in life science a "mystical" propensity – or a (privatively) "aesthetic" one – that deeply misprises (as irrational) the *interpretive* idea of science that was being developed by its most brilliant eighteenth-century expositors – Buffon, Daubenton, Diderot, Camper, Goethe and Herder. See Daston and Galison 2007.

17 "The formative drive is not the *cause* of this leap [from inorganic to organic], but rather its *expression*" (McLaughlin 1982, 364). I share the view of Robert Richards (2002, 221–222) that the relation in Blumenbach in fact tended to flow in the other direction, even if Blumenbach's metaphysical preferences inclined him to want to see it as McLaughlin reconstructs.

18 See Duchesneau 1985. This, of course, becomes the Achilles heel of "vitalism" in historical retrospect.

19 Christoph Girtanner would pick this up explicitly. See Girtanner 1796.

References

Aulie, R. 1961. Caspar Friedrich Wolff and his "Theoria Generationis," 1759. *Journal of the History of Medicine*, *16*: 124–144.

Baer, K. von. 1828. *Über Entwicklungsgeschichte der Thiere*. Königsberg: Bornträger.

Blumenbach, J. F. 1779. *Handbuch der Naturgeschichte, Erster Teil*. Göttingen: Dieterich.

Blumenbach, J. F. 1781. *Über den Bildungstrieb und das Zeugungsgeschäfte*. Stuttgart: G. Fischer, 1971.

Blumenbach, J. F. 1782. *Handbuch der Naturgeschichte*, 2nd edition. Göttingen: Dieterich.

Blumenbach, J. F. 1785. *Medicinische Bibliothek 2*. Göttingen: Dieterich.

Blumenbach, J. F. 1788. *Handbuch der Naturgeschichte*. Göttingen: Dieterich.

Blumenbach, J. F. 1789. *Über den Bildungstrieb*. Göttingen: Dieterich.

Blumenbach, J. F. 1791. *Handbuch der Naturgeschichte*. Göttingen: Dieterich.

Blumenbach, J. F. 1797. *Handbuch der Naturgeschichte*. Göttingen: Dieterich.

Bonnet, C. 1776. *Considérations sur les corps organisés*, 3rd edition. Amsterdam: M. M. Rey.

Bonnet, C. 1782. *Contemplation de la nature*, 3rd, revised and expanded edition. Hamburg: Virchaux.

Breidbach, O. 1995. Die Geburt des Lebendigen – Embryogenese der Formen oder Embryologie der Natur? – Anmerkungen zum Bezug von Embryologie und Organismustheorien vor 1800. *Biologisches Zentralblatt*, *114*: 191–199.

Buffon, G. L. L. Comte de. 1749. De la manière d'étudier et de traiter l'histoire naturelle. Histoire et théorie de la terre (vol I) and Histoire générale des animaux. Histoire naturelle de l'homme (vol II). In *Histoire Naturelle, générale et particulière, avec la description du cabinet du roi*, 44 Volumes. Paris: Imprimerie Royale, 1749–1804.

Cherni, A. 1998. *Épistémologie de la transparence: sur l'embryologie de A. Von Haller*. Paris: Vrin.

Clarke, R. 1942. Herder's concept of "*Kraft.*" *PMLA*, *57*: 737–752.

Daston, L. and Galison, P. 2007. *Objectivity*. New York: Zone.

Duchesneau, F. 1985. Vitalism in late eighteenth-century physiology: The cases of Barthez, Blumenbach and John Hunter. In W. F. Bynum and R. Porter (eds.) *William Hunter and the Eighteenth-Century Medical World* (pp. 259–295). Cambridge: Cambridge University Press.

Duchesneau, F. 2006. Essential force and formative force: Models for epigenesis in the 18th century. In B. Feltz, M. Commelinck and P. Goujon (eds.) *Self-Organization and Emergence in the Life Sciences* (pp. 171–186). Dordrecht: Springer.

Dupont, J.-C. 2007. Pre-Kantian revival of epigenesis: Caspar Friedrich Wolff's *De formatione intestinorum* (1768–1769). In P. Huneman (ed.) *Understanding Purpose: Collected Essays on Kant and Philosophy of Biology* (pp. 37–49). Rochester, NY: University of Rochester Press/North American Kant Society Studies in Philosophy.

Gaissinovich, A. E. 1968. Le rôle du Newtonianisme dans la renaissance des idées épigénetiques en embryologie du XVIIIe siècle. In *Actes du XIe Congrès International d'Histoire des Sciences*, *5*: 105–110.

Gaissinovich, A. E. 1990. C. F. Wolff on variability and heredity. *History and Philosophy of the Life Sciences*, *12*, 179–201.

Girtanner, C. 1796. *Über das Kantische Prinzip für die Naturgeschichte*. Göttingen: Vandenhoeck & Ruprecht.

Goethe, J. W. von. 1988. The formative impulse. In *Scientific Studies* (pp. 35–36). New York: Suhrkamp.

Goy, I. 2014. Epigenetic theories: Caspar Friedrich Wolff and Immanuel Kant. In I. Goy and E. Watkins (eds.) *Kant's Theory of Biology* (pp. 43–60). Berlin/Boston, MA: De Gruyter.

Goy, I. and E. Watkins (eds.). 2014. *Kant's Theory of Biology*. Berlin/Boston, MA: De Gruyter.

Hall, T. S. 1968. On biological analogs of Newtonian paradigms. *Philosophy of Science*, *35*: 6–27.

Haller, A. von. 1751. *Réflexions sur le système de la génération de M. de Buffon*. Geneva: Barrillot.

Haller, A. von. 1758. *Sur la formation du coeur dans le poulet: sur l'oeil, sur la structure du jaune, etc*. 2 vols; Lausanne: Bousquet.

Haller, A. von. 1760. Review of Theoria Generationis. *Göttingische Anzeigen von gelehrten Sachen*: 1226–1231.

Haller, A. von. 1766. *Elementa physiologiae corporis humani*, 8 vols. Lausanne: Bousquet, 1757–1766; German translation: *Herrn Albrecht von Hallers Anfangsgründe der Phisiologie des menschlichen Körpers*, trans. Johann Samuel Halle, 8 vols. Berlin: Voss, 1762–1776.

Haller, A. von. 1981. Buffon on hypotheses: The Haller preface to the German translation of the Histoire Naturelle 1750, trans. Phillip Sloan, in J. Lyon & P. Sloan (eds. and trans.) *From Natural History to the History of Nature: Readings in Buffon and His Critics* (pp. 295–310). Notre Dame, IN/London: University of Notre Dame Press.

Harvey, W. 1651. *On Generation*. Reprint: Ann Arbor, MI: Edwards, 1943.

Herder, J. G. 1775. Über die dem Menschen angeborene Lüge. In B. Suphan (ed.) *Sämmtliche Werke* (vol. 9, pp. 536–540), 33 vols. Berlin: Weidmann, 1880–1899.

Herder, J. G. 1778. Vom Erkennen und Empfinden der menschlichen Seele: Bemerkungen und Träume. In *Sämmtliche Werke* (vol. 8, pp. 165–262).

Herder, J. G. 1784. Ideen zur Philosophie der Geschichte der Menschheit. *Sämtliche Werke*, vol. 13: 1–439; 14: 3–493.

Herder, J. G. 1943. *God: Some Conversations*, trans. F. Burkhardt. New York: Veritas.

Herrlinger, R. 1959. C. F. Wolff's "Theoria generationis" (1759). *Zeitschrift für Anatomie und Entwicklungsgeschichte, 121*: 245–270.

Herrlinger, R. 1966. "Vorwort," to C. F. Wolff, *Theorie von der Generation in zwei Abhandlungen erklärt und bewiesen* (1764). *Theoria generationis* (1759), 5–28. Reprint: Hildesheim, Olms.

Hoquet, T. 2005. *Buffon: histoire naturelle et philosophie*. Paris: Champion.

Huneman, P. 2007a. Reflexive judgment and Wolffian embryology: Kant's shift between the first and the third critiques. In P. Huneman (ed.) *Understanding Purpose: Collected Essays on Kant and Philosophy of Biology* (pp. 75–100). Rochester, NY: University of Rochester Press/North American Kant Society Studies in Philosophy.

Huneman, P. (ed.). 2007b. *Understanding Purpose: Collected Essays on Kant and Philosophy of Biology*. University of Rochester Press/North American Kant Society Studies in Philosophy.

Irmscher, H.-D. 1973. Grundfragen der Geschichtsphilosophie. In Johann Maltusch 1971 (ed.) *Bückeburger Gespräche über Johann Gottfried Herder* (pp. 17–57). Bückeburg: Grimme.

Irmscher, H.-D. 1987. Die geschichtsphilosophische Kontroverse zwischen Kant und Herder. In *Hamann–Kant–Herder. Acta des vierten Internationalen Hamann-Kolloquiums im Herder-Institut zu Marburg/Lahn 1985* (pp. 111–192). Frankfurt: Regensburger Beiträge zur deutschen Sprach- und Literaturwissenschaft, Reihe B, 34.

Jahn, I. 1998–1999. Wer regte Caspar Friedrich Wolff (1734–1794) zu seiner Dissertation "Theoria generationis" an? *Philosophia Scientiae*, Cahier Spécial (2): 35–54.

Jahn, I. 1999. Maupertuis zwischen Präformations- und Epigenesis-theorie. Sein Beitrag zu biologischen Fragen des 18. Jahrhunderts. In H. Hecht (ed.) *Pierre Louis Moreau de Maupertuis: Eine Bilanz nach 300 Jahren* (pp. 111–192). Berlin: Spitz.

Jardine, N. 2000. *The Scenes of Inquiry: On the Reality of Questions in the Sciences*. 2nd ed. Oxford: Clarendon.

Kant, I. 1763. *Der Einzig Mögliche Beweisgrund zu einer Demonstration des Daseins Gottes*. AA:2: 63–164.

Kant, I. 1766. *Träume eines Geistersehers, erläutert durch Träume der Metaphysik*. AA:2: 315–384.

Kant, I. 1785. *Recensionen von J. G. Herders Ideen zur Philosophie der Geschichte der Menschheit*. Theil 1. 2., AA:8: 43–66.

Kant, I. 1786. *Metaphysische Anfangsgründe der Naturwissenschaft*. AA:4: 465–566.

Kant, I. 1788. *Über den Gebrauch teleologischer Principien in der Philosophie*. AA:8: 157–184.

Kant, I. 1790. *Kritik der Urtheilskraft*. AA:5: 165–486.

Kant, I. 1901–Present. *Gesammelte Schriften Herausgegeben von der Preußischen Akademie der Wissenschaften* (Vols. 1–29, to date). Berlin: De Gruyter.

Kant, I. 1922. *Briefwechsel*, AA:11.

Karolyi, L. V. 1971. "Vorwort" to Blumenbach. In *Über den Bildungstrieb und das Zeugungsgeschäfte* (pp. v–xx). Stuttgart: G. Fischer.

Lenoir, T. 1980. Kant, Blumenbach, and vital materialism in German biology. *Isis, 71*: 77–108.

Locke, J. 1689. *An Essay Concerning Human Understanding*. Indianapolis, IN: Hackett, 1996.

Lukina, T. A. 1975. Caspar Friedrich Wolff und die Petersburger Akademie der Wissenschaften. *Acta historica Leopoldina 9*: 411–425.

Maupertuis, P. M. de. 1752. *Lettre sur le progrès des sciences*.

Maupertuis, P. M. de. 1756. *Système de la nature*. In *Essai de cosmologie, Système de la Nature, Réponse aux Objections de M. Diderot*, ed. F. Azouvi. Paris: Vrin, 1984.

McLaughlin P. 1982. Blumenbach und der Bildungstrieb: zum Verhältnis von epigenetischer Embryologie und typologischem Artbegriff. *Medizinhistorisches Journal, 17*: 357–372.

McLaughlin, P. 1990. *Kant's Critique of Teleology in Biological Explanation: Antinomy and Teleology*. Lewiston, NY: Mellen.

Medicus, F. C. 1774. *Von der Lebenskraft*. Mannheim: n.p.

Mensch, J. 2013. *Kant's Organism*. Chicago, IL: University of Chicago Press.

Mocek, R. 1995. Caspar Friedrich Wolffs Epigenesis-Konzept – ein Problem im Wandel der Zeit. *Biologisches Zentralblatt, 114*: 179–190.

Niewöhner, F. and J.-L. Seban (eds.). 2001. *Die Seele der Tiere*. Wiesbaden: Harrassowitz.

Nisbet, H. B. 1970. *Herder and the Philosophy and History of Science*. Cambridge: Modern Humanities Research Association.

Palti, E. 1999. The "metaphor of life": Herder's philosophy of history and uneven developments in late eighteenth-century natural sciences. *History and Theory, 38*: 322–347.

Proß, W. 1987. [Commentary on] "Fragmente über Wolff, Baumgarten und Leibniz." In J. G. Herder, *Werke, Bd. II: Herder und die Anthropologie der Aufklärung*, ed. W. Proß (pp. 847–865). Darmstadt: Wissenschaftliche Buchgesellschaft.

Reil, J. C. 1795. *Von der Lebenskraft, Eingeleitet von Herausgeber*, vol. 2, ed. Karl Sudhoff. Leipzig: Barth.

Reill, P. H. 2005. *Vitalizing Nature in the Enlightenment*. Berkeley/Los Angeles, CA: University of California Press.

Richards, R. 2002. *The Romantic Conception of Life: Science and Philosophy in the Age of Goethe*. Chicago, IL/London: University of Chicago Press.

Roe, S. 1979. Rationalism and embryology: Caspar Friedrich Wolff's theory of epigenesis. *Journal of the History of Biology, 12*: 1–43.

Roe, S. 1981. *Matter, Life and Generation: Eighteenth-Century Embryology and the Haller-Wolff Debate*. Cambridge: Cambridge University Press.

Schelling, F. 1798. *Von der Weltseele: eine Hypothese der höheren Physik zur Erklärung des allgemeinen Organismus*. Hamburg: Perthes.

Schuster, J. 1941. Der Streit um die Erkenntnis des organischen Werdens im Lichte der Briefe C. F. Wolffs an A. von Haller. *Sudhoffs Archiv, 34*: 196–218.

Sloan, P. 2002. Preforming the categories: Kant and eighteenth-century generation theory. *Journal of the History of Philosophy, 2*: 229–253.

Smith, J. E. H. (ed.). 2006. *The Problem of Animal Generation in Early Modern Philosophy*. Cambridge, New York: Cambridge University Press.

Sonntag, O. (ed.). 1983. *The Correspondence between Albrecht von Haller and Charles Bonnet*. Bern, Stuttgart, Vienna: Huber.

Stolpe, H. 1964. Herder und die Ansätze einer naturgeschichtlichen Entwicklungslehre im 18. Jahrhundert. In *Neue Beiträge zur Literatur der Aufklärung* (pp. 289–316, 454–468). Berlin: Rütten & Loening.

Tetens, J. N. 1777. *Philosophische Versuche über die menschliche Natur und ihre Entwickelung*, 2 vols. Leipzig: Weidman.

Uschmann, G. 1955. *Caspar Friedrich Wolff: Ein Pionier der modernen Embryologie*. Leipzig/Jena: Urania.

Wenzel, M. 1990. Die Anthropologie Johann Gottfried Herders und das klassische Humanitätsideal. In G. Mann and F. Dumont (eds.) *Die Natur des Menschen: Probleme der Physischen Anthropologie und Rassenkunde (1750–1850)* (Soemmerring Forschungen 6) (pp. 137–167). Stuttgart: G. Fischer.

Wolfe, C. T. 2014. On the role of Newtonian analogies in eighteenth-century life science: Vitalism and provisionally inexplicable explicative devices. In Z. Biener and E. Schliesser (eds.) *Newton and Empiricism* (pp. 223–261). Oxford: Oxford University Press.

Wolff, C. F. 1759. *Theoria generationis* (1759), reprinted in *Theorie von der Generation in zwei Abhandlungen erklärt und bewiesen* (1764). *Theoria generationis* (1759), ed. and intro. Robert Herrlinger. Stuttgart: G Fischer, 1966.

Wolff, C. F. 1764. *Theorie von der Generation in zwei Abhandlungen erklärt und bewiesen* (1764), reprinted in: *Theoria generationis* (1759), ed. and intro. Robert Herrlinger. Stuttgart: G Fischer, 1966.

Wolff, C. F. 1774. *Theoria generationis*, ed. Philip Meckel. Halle: Hendel.

Wolff, C. F. 1789a. Von der eigenthümlichen und wesentlichen Kraft der vegetabilischen sowohl als auch der animalischen Substanz. In *Zwo Abhandlungen über die Nutritionskraft ... nebst einer fernen Erläuterung eben derselben Materie von C. F. Wolff.* St. Petersburg: Royal Academy of Sciences.

Wolff, C. F. (ed.). 1789b. *Zwo Abhandlungen über die Nutritionskraft welche von der Kayserlichen Academie der Wissenschaften in Saint Petersburg den Preis getheilt erhalten haben. Nebst einer fernen Erläuterung eben derselben Materie von Caspar Friedrich Wolff.* St. Peterburg: Kayserlichen Academie der Wissenschaften.

Wolff, C. F. 1812. *Über die Bildung des Darmkanals im bebrüteten Hühnchen*, trans. Meckel. Halle: Benger.

Wolff, C. F. 1896. *Theoria generationis*, 2 vols., trans. P. Samassa. Leipzig: Engelmann.

Wolff, C. F. 2003. *De formatione intestinorum = La formation des intestins: 1768–1769*, ed. and trans. Jean-Claude Dupont. Turnhout: Brepols.

Wright J. P. and P. Potter (eds.). 2000. *Psyche and Soma: Physicians and Metaphysicians on the Mind-Body Problem from Antiquity to Enlightenment.* Oxford: Clarendon.

Zammito, J. H. 1998. "Method" versus "manner"? Kant's critique of Herder's *Ideen* in the light of the epoch of science, 1790–1820. In H. Adler and W. Koepke (eds.) *Herder Jahrbuch/Herder Yearbook, 1998* (pp. 1–26). Stuttgart: Metzler.

Zammito, J. H. 2001. Epigenesis: Concept and metaphor in Herder's *Ideen.* In R. Otto and J. H. Zammito (eds.) *Vom Selbstdenken: Aufklärung und Aufklärungskritik in Herders "Ideen zur Philosophie der Geschichte der Menschheit"* (pp. 129–144). Heidelberg: Synchron.

Zammito, J. H. 2003. "This inscrutable principle of an original organization": Epigenesis and "looseness of fit" in Kant's philosophy of science. *Studies in History and Philosophy of Science, 34*: 73–109.

Zammito, J. H. 2006. Kant's early views on epigenesis: The role of Maupertuis. In J. E. H. Smith (ed.) *The Problem of Animal Generation in Early Modern Philosophy* (pp. 317–354). Cambridge: Cambridge University Press.

Zöller, G. 1988. Kant on the generation of metaphysical knowledge. In H. Oberer and G. Seel (eds.) *Kant: Analysen – Probleme – Kritik* (pp. 71–90). Würzburg: Königshausen & Neumann.

Part II

Organism and organization

5 Senebier and the advent of general physiology

François Duchesneau

Introduction

Jean Senebier (1742–1809) played a significant part in the advent of general physiology, a few decades before the term and concept became of use to characterize a fundamental approach in biology. In that perspective it is probably worth taking into account Senebier's researches on the chemistry of gas exchanges as the main spring of organic life in plants. However, this viewpoint may appear to be too restrictive for describing the place Senebier deserves to hold in the history of general physiology. One should also consider, besides vegetal physiology, two other sides of his work that are not to be underrated: on the one hand, the amplifying interpretation he grafts to the researches of his friend Lazzaro Spallanzani on animal physiology, in particular concerning respiration; on the other hand, the methodology he sets on stage to indicate the orientation according to which the new physiology should be able to display laws ruling over the system of organic and living nature. This latter methodological contribution should also be correlated with some of the developments in the second edition of his *Essai sur l'art d'observer et de faire des expériences* (1802).[1] But here I shall more particularly address the following points: (1) the program for general physiology that emanates from Senebier's work on the functioning of plants; and (2) the theoretical extension he brought onto Spallanzani's experimental researches in animal physiology. In both cases, I shall be concerned with the methodological framework he proposed for advancing a systemic representation of vital processes.

Physiology and the functioning of plants

I shall first refer to Senebier's *Physiologie végétale* (1800) for its dominant methodological and theoretical features. The author wants to achieve in that work a "systematic ordering" (*ordre systématique*) of the most complete set of observations and experiments in anatomy and physiology with "a constant application of physics and chemistry to the study of plants" (Senebier 1800, I, 11). Analysis is thus called upon to focus on the organic composition of plants and the chemical and physical processes they undergo: these arise either from the determining influence of factors of the external milieu or from the interconnection of inner operations in the various parts. Senebier professes that the general phenomena of nature, namely the general phenomena of the physical and chemical order, averred through constant observations of empirical data, express themselves through the specific processes of the organic order which they result into (Senebier 1800, I, 24). What characterizes the proposed approach is its analytic reductionism. The organic being that observation first presents us with its complexity must be decomposed into its constitutive parts, but the enumeration and anatomical description of these structures, however extended, cannot exempt us from focusing on their effects, relations and functions, analyzed into processes and variations of processes.

> One must add to the detailed anatomy of plant organs, a careful study of their effects, relations, importance of ends, and reality of achievements, and this supposes a reflective observation of the phenomena generated by these organs' reciprocal action, or by that of their parts on one another, together with a careful investigation of the way this action exerts itself under various circumstances.
>
> (Senebier 1800, I, 19–20)

Senebier does not shrink from resorting to mechanist analogies to describe this analytic approach to processes that, through combining their effects within the organic structure, yield the development, relative conservation and reproduction of those "complex machines" (*machines compliquées*) (Senebier 1800, I, 22), which plants form, as do all organized bodies. At the same time, he holds that these physical and chemical processes occur by and through the interaction of organic components and form the causal sequences out of which living phenomena arise. The challenge for the analytic way consists not only in unveiling those organic causal sequences but also in understanding how such causal sequences can combine and harmonize so as to produce the integration of processes in the life of the living being as a constantly changing whole. In these conditions, anatomy, even microscopic and comparative anatomy, does not suffice for revealing:

> the secret combination of organs, the force that moves them, the laboratory of the juices they prepare, which renders them apt to assimilate and produce so many diverse matters, the hub that connects so many contrary elements, and finally the means that harmonically develop and preserve this whole made out of adverse substances, by making a host of motions issuing from thousands of particular causes collaborate to maintain the qualities of each species. Such is however the chain of causes and effect we ought to conceive, or rather to seek in the plants themselves.
>
> (Senebier 1800, I, 22)

The physiology that Senebier conceives appears as a science of relations whereas anatomy is defined as bearing only on the organic components considered per se.[2] Progress in this way of doing physiology consists in fostering a specific mode of analysis. This evolution may thus be considered as promoting the type of experimental analysis Spallanzani had illustrated, and forming the methodological kingpin of a science of vital functioning that is fundamentally based on observation data. But, if this empirical basis constitutes the ground-piece or starting-block for such a type of analysis, one also needs to resort to the most appropriate conceptual practices for making sense of the investigated facts of experience. The challenge is to infer from these facts causes and laws that would own the highest epistemic probability and truly account for the order of phenomena concerned. In the context of Senebier's researches, the aim is thus to start from accessible experimental data and discover the causes and laws that would explain the emergence and order of production of such complex organic and vital phenomena.

Senebier undertakes to show how, in the case of plants, this physiology is to be built. The best illustration to this purpose is provided in the sections of the chapter devoted to light: these sections relate to the assimilation of carbonic acid (carbon dioxide, CO_2) and to the role of green parts in the dynamic of the physical-chemical processes involved.[3] In this instance, Senebier's observations and interpretive hypotheses belong to a trend that historians of life sciences have well identified:[4] this trend progressively developed into the photosynthesis theory and the explanation of the chlorophyll function. In this discovery process, Senebier holds a particular place. Charles Bonnet, his mentor in the life sciences, had observed that leaves of green plants emitted gas bubbles when immersed in water. Joseph Priestley, later on, had succeeded in isolating this gas and determining his necessary role in maintaining animal life. And Antoine de Lavoisier had established the replacement of oxygen by carbon dioxide in respiration. But it was Jan Ingenhouz, publishing his *Experiments upon Vegetables, Discovering Their Great Power of Purifying the Common Air in the Sunshine and of Injuring It in the Shade and at Night* (1779), who set the notion of "balance of life" by establishing the determinate and reciprocal relation in gas exchanges that linked animal with vegetal physiology: according to his hypothesis, green plants absorb atmospheric CO_2 during the night and release oxygen during the day under the influence of sunlight. In his *Mémoires physico-chimiques sur l'influence de la lumière solaire pour modifier les Êtres des trois règnes de la nature* (Senebier 1782), completed by his *Expériences sur l'influence de la lumière solaire dans la végétation* (Senebier 1788), Senebier resumes and corrects Ingenhouz's hypothesis by establishing that the carbonic acid gas reaches the vegetal organism not only through extraction from atmospheric air but also through radicular absorption from a state of dissolution in water, notably in water flowing from soils composed of decaying organisms. Above all, Senebier is interested in the chemical reactions resulting from transformed CO_2 and metabolizing the various organic constituents of plants out of carbonated substances. He correlatively shows the diversity and complementarity of gas exchanges produced in leaves under the influence of solar radiations and resulting in the exhalation of oxygen and of a given amount of untransformed CO_2. The metabolic processes of nutrition in plants are made the object of the most systematic deciphering possible in terms of chemical reactions. This analysis is then completed by a synthesis of the relations occurring between physical-chemical factors, and it is made to represent the trophic basis on and according to which the functional economy of plants is built up. Later on, this integrative synthesis of analytic relations will tend to similarly include the functional economy of animals, or, more exactly, this part of the project, previously only sketched, will develop, in parallel with Senebier's contributions to plant physiology, in various essays accompanying or prolonging the diffusion of Spallanzani's experimental researches in animal physiology.

In vegetal physiology, Senebier's demonstration frames up as follows: the matter is about how carbon is provided to the plant so as to enter into various combinations with its structures and the products of its functional activity. The hypothesis to corroborate is that this carbon, an elementary substance, results from decomposition of the carbonic acid gas; the latter would come about by the "act of vegetation" (*acte de végétation*) achieved in the parenchyma of leaves by the effect of solar radiations (Senebier 1800, III, 157). The carbon thus extracted from CO_2 of either atmospheric or terrestrial origin would then combine in various parts with other principles so as to produce "organic juices" (*sucs végétaux*). The continuous supply of this dissolved gas in organic fluids grounds and enables vital functioning in plants.

> Earth, water, and air provide the elements and vehicle of carbonic acid in plants; this acid appears as the most important nutriment they receive, which can be easily inferred from the quantity of carbon they contain. Carbonic acid also plays a very important role in vegetal economy; it penetrates with water in the roots which suck it. It rises with the sap that it contributes to form up into the leaves, where it is decomposed by the action of light. Carbonic acid spread in the vapours of the atmosphere is deposited with the water the latter forms on the leaves which absorb it continuously: this provides plants with a constant, abundant and uniform nutriment, more precisely carbon and oxygen, which they do not seem to be able to receive otherwise.
>
> (Senebier 1800, III, 165)

The demonstrative strategy set up consists on the one hand in determining experimentally: (1) the sources of supply of the CO_2 obtained by dissolution in water: (2) the absorption of the latter by the peripheral organs: roots and leaves; (3) the modalities of its decomposition by the action of light without emission of caloric; (4) the corresponding exhalation of oxygen by the leaves; and (5) the combination of carbon along specific affinities in the various solid and liquid components of the vegetal organism. In all cases, experiments are devised so as to fit the diversification and variation of parameters and facilitate quantifying the processes and products resulting from vegetative activity. On the other hand, a progressive generalization of cases conformable to the experimental pattern takes place. Thus, if initially the cases of carbon dioxide absorption and oxygen exhalation considered were those that leaves achieve when immersed in common water or water with CO_2 added, with or without contact with the external atmosphere, one then passes on to the consideration of absorptions achieved in atmospheric air by the dissolution of dew, which plants suck.

In the systematic analysis of empirical data fitting his hypothesis, Senebier annexes the experiments of other researchers, notably those of Nicolas Théodore de Saussure.[5] Finally, Senebier submits his hypothesis and the systematic experimental analysis that goes with it to crosschecking in light of the difficulties and objections raised by other experimenting naturalists.[6] As he mentions about an issue raised by Spallanzani,

> this observation surprised me, and I proceeded instantly to examine this fact in order to abandon my theory, if I were unable to find the cause of it, or to delve deeper into this matter, if I succeeded in clarifying this difficulty.
>
> (Senebier 1800, III, 228)

According to Spallanzani, perennial plants immersed in water totally lacking carbonic acid would have been observed to keep emitting oxygen. Through a remarkable series of experiments, Senebier succeeds in establishing that the oxygen thus produced had come from CO_2 previously stocked in a dissolved state in the juices of the leaves' parenchyma (Senebier 1800, III, 249). Senebier replies similarly, by means of quantitative measurements, to the objection that, according to Felice Fontana's hypothesis, these plants only give back oxygen that was contained in the water. He also develops in detail a possible reply to Jean-Henri Hassenfratz's objections. The latter had objected to Senebier's statement that assimilated carbon is proportional to absorbed CO_2, because the quantity of carbon contained in the organism varies but slightly in comparison with the presumed ratios of absorption and decomposition of CO_2. In his reply, Senebier argues that carbon assimilation holds a determined relationship to the organization concerned, so that any excess of carbonic acid gets eliminated, along with the evacuation of oxygen into the atmosphere. On the other hand, the source of supply in carbon cannot be provided but by the decomposition of CO_2, if the case of aquatic or immersed terrestrial plants is well taken into account. Faced with an adverse hypothesis that implied the decomposition of water rather than of CO_2, he deems it appropriate to appeal to the greater probability resulting from the experimental inferences on the action of light he had drawn from empirically confirmed decomposition of carbonic acid (Senebier 1800, III, 255). Finally, the presumed direct assimilation of carbon dissolved in water should also be excluded for factual reasons. Senebier's ultimate aim is to present the general law of absorption and decomposition underpinning oxygen production by the green leaves under the action of sunlight, while taking into account the greatest possible number of factors that could modify the quantitative relations involved. If one has to admit a constant relationship between absorbed carbonic acid and exhaled oxygen, this relationship shall be consequently modified in determinate ways according to the parameters ruling over the implied physical-chemical processes.[7]

Senebier develops in analogous manner an analytic conception of germination. The need for air for seeds to germinate is analytically reduced to the need for combinations to occur that imply oxygen and result into the formation of CO_2 in specific conditions of humidity and warmth and outside of the presence of sunlight. For that light would stop fermentation, with which germination owns the closest affinity, and trigger nutrition processes like those of plants once they have emerged from the buried seed.[8]

These analyses and the experimental data they comprise belong to a systematic approach of phenomena that Senebier refers to at the end of his *Physiologie végétale*. Several features are then emphasized concerning the methodology of a science that is judged "not to have left the cradle yet" (*encore au berceau*) (Senebier 1800, V, 230). One of the instructions he gives concerns the need to revise the comparison of plants with animals. On this subject, analogy should give way to generic determinations applying to the complete set of living organisms, along with the appraisal of significant differences in structural and functional modalities for each type of organization. Distinctions should be taken into account about these organizations' solid and fluid components, their modes of composition, their ways of framing up, the processes they undergo, but also their functional dispositions to motility, irritability, respiration, nutrition, growth, and reproduction.[9] This does not mean however that we should renounce forming syntheses concerning the functioning of organisms, especially since the access to general physiology is there at stake. Senebier underlines:

> I acknowledge indeed that the economy of plants is up to a given point reducible to animal economy, but would this not be because this economy fits into the plan of the universe with relationships quite similar to those of animals?
>
> (Senebier 1800, V, 201)

Thus, since physiology is a science of relations, these express or translate the infinitely variable organization of life forms, and they imply generic affinities as well as structural and functional differences that may be detailed in multiple ways. Along another aspect of vital organizations, innumerable relationships reflect determinations toward inner change that result from external factors, or physical and chemical interactions between the components and products of organisms. To match these varieties in organization and reciprocal determinations between organs, processes and contextual factors, analysis will resort to means provided by physics and chemistry, for:

> one needs not only to know the essential properties of these [organic] bodies, but also the effects resulting from the relations they have with the bodies and substances surrounding them, because we must constantly discover their influence, assign their causes, measure their energy and discern their role.
>
> (Senebier 1800, V, 224)

In this perspective, modern chemistry, which Lavoisier had undertaken to codify, affords a new means "for penetrating into the laboratories of nature" (Senebier 1800, V, 225) that organisms form, whether plants or animals, for their inherent and constitutive relationships as well as for the emergent relationships they generate.[10] On the other hand, analyzing the sets of constitutive relations for the various organizations will entail taking into account the diachronic dimension of organizations and processes in the formation and development of organisms, and this analysis will also need to consider the structural and functional variants in organization that differentiate species of individuals. For instance:

> [The *desiderata* of vegetal physiology] will be founded on the natural or constrained relationships between the various plants and the different substances suitable to act on them, when considered in their states of health, illness or destruction; that is to say in all moments of their life and in all circumstances, natural or extraordinary, in which they can be found. And these desiderata will be founded on the comparative anatomy and physiology of plants with one another, as well as on the effects resulting from the diverse organization of each in particular, and on the similarities and differences they display.
>
> (Senebier, 1800, V, 233–234)

The means for implementing this method boil down to the main ones that observation and experience represent: in both cases, the aim is to comprehend the constitutive relations of organization or those issuing therefrom. Experience in particular reveals determinate relations that can be varied and tested in multiple ways with distinct organisms. Beyond that, interpreting those phenomena requires resorting to analogies and hypotheses, subject nevertheless to strict empirical controls (Senebier 1800, V, 227–230). Can these patterns for experimental practice and demonstrative approach apply by extension and analogy to animal physiology, under the common heading of general physiology?

Extension of patterns to animal physiology

Senebier had views about the development of animal physiology similar to those which his works on plant physiology had exposed. The model for physiology he started from was that of Haller,[11] coupled with considerations inspired by Bonnet's theoretical conceptions.[12] But the model that progressively determined the methodological orientation he adhered to and supported came from Spallanzani, whose main experimental memoirs he had translated and edited in the mature period of his career.[13] Two features of Senebier's contributions in that domain deserve consideration. While diffusing the results of Spallanzani's experimental researches, the Genevan naturalist set to himself the task of drawing out their theoretical framework and inferring therefrom guidelines for a more extensive and systematic scientific practice. Previous editions of works by Spallanzani, in particular the three volumes published in Pavia in 1787, had already included pieces by Senebier fulfilling that twofold objective, for instance the *Introduction du traducteur, dans laquelle on fait connoître la plupart des Découvertes microscopiques faites dans les trois Règnes de la Nature, avec leur influence sur la perfection de l'Esprit humain*; (Spallanzani 1787, I, i–cxiv) the *Considérations sur la méthode suivie par Monsieur l'Abbé Spallanzani dans ses expériences sur la digestion* (Spallanzani 1787, II, 315–393); and the *Ébauche de l'histoire des Êtres organisés avant leur fécondation* (Spallanzani 1787, III, i–lxxxviii). But if one wishes to grasp what in Senebier's approach corresponds exactly to the purpose of grounding animal physiology on bases similar to those of the vegetal physiology he had achieved according to his own plans, one should turn more specifically to the pieces accompanying the edition of the *Rapports de l'air avec les Êtres organisés* (1807). Describing the contents of the *Rapports*, Senebier underlines "the slow progression of analysis in its aim at reaching capital truths" (*la marche lente de l'analyse pour arriver aux vérités capitales*) (Senebier 1807, I, xii) and the fact that physiologists tend to occult in their statements the lengthy processes through which they have reached the discoveries they are credited for. Spallanzani is however presented as the protagonist of a new analytic method that will give access to the most fundamental theories on the organic realm.

> These memoirs will present a new way of studying animals, new relationships to grasp to write their history, original chapters to complement their physiology, and grand views to add to those that form that science. Reading this work, one will sense that we did not know, to say the least, the relation of air with the animal realm. One will see that Spallanzani opens a new era to naturalists and physicians, provides data for explaining various unresolved phenomena, and leads to search for those consequential facts that may unveil the mysteries of assimilation and those of the permanency of laws of nature in the animal economy.
>
> (Senebier 1807, I, xii–xiii)

Toward a new physiology that may yield the most general views possible on animal organization, Senebier's *Considérations générales sur la respiration* (Senebier 1807, II, 260–344), as well as the *Mémoire sur les animaux léthargiques*, mark important steps (Senebier 1807, II, 345–402).

At the beginning of the former memoir, Senebier states that "constant and uniform relations" link all organized beings with atmospheric air. Whereas infinite variety seems to characterize "their constitution, organization, destination, habits, as well as their different ways of existing at the various period of their life, including even those that follow their death" (Senebier 1807, II, 260), these relations, which may be analytically reduced to general facts, express a determining condition, probably the most essential, for the life and destruction of individuals, as well as of species, and even of the entirety of life forms considered both synchronically and diachronically. The same functional disposition in terms of basic chemical exchanges shows up in the constantly rearranged alternation between storing combustible substances in the organism and breaking up bonds and affinities issuing from combustion. Analysis must spell out the generality of these basic relationships, namely those of respiration and its vegetal equivalent, down to the multiple means by which they are achieved. Because of these relationships which represent a fundamental law of animalization and vitalization, Senebier rejects appeals to any concept of vital force. The term "vital force" may be held to designate an unknowable entity and therefore it should be banished from the explanation of phenomena, unless it is used for referring to a complex set of operations arising from the specific organization of living beings according to which their basic chemical processes are regulated.

> The life of the animal greatly fosters the absorption of oxygen gas; experience has shown that a living frog deprived of its lungs absorbs more oxygen than one that has just been killed. It is easily felt that this difference depends on the motion of blood, on the reaction of organs to each other, on the combinations that occur, on the decompositions that cause them, on the combination of oxygen with carbon resulting from these, and not on this vital force which is but a name given to an unknown entity and cannot offer any idea, unless it should represent this reciprocal action of organs; whatever be the case, in the living animal, this combination of oxygen with carbon operates for the sake of its conservation, and in the dead animal for the sake of its destruction.
>
> (Senebier 1807, II, 294–295)

The analytic approach Senebier intends to apply for extending Spallanzani's observations would bear on the atmospheric gases and their combinations with carbon or dissociations from it in the various apparatuses, organs and structural components of animal organisms, including elementary fibers, in particular muscular fibers. This is what comes out of the suggested hypothesis to account for the nonsaturation of the solid and fluid parts or organisms by the oxygen absorbed by the inner and outer surfaces of the body and by the blood. This assimilation mechanism is identified as the determining condition for irritability and, so, as the agent par excellence of the vital dynamic. Senebier figures out that this process is paired with the production of carbonic acid and its rejection into external gaseous milieus: the link between the two sets of processes should be conceived as the interplay of inverse affinities conjoined in the same organic locations. Senebier thus suggests:

> Is it not probable that animals which all reject carbonic acid through the lungs and the surface of their body form it by the property that muscles and skin would have to seize the caloric of the oxygen gas? Then oxygen would seize carbon, and carbonic acid whose affinity with the muscle is lesser than that of oxygen would leave it while caloric withdrawn from oxygen would favor a new union of this oxygen with the muscle or with its carbon, or maybe with both; but as there would still be some stored in the muscle and

> skin, without its being entirely decomposed, it would be left to provide for a while the carbonic acid produced by animals placed in mephitic gases.
> (Senebier 1807, II, 303–304)

This hypothesis would not suffice to provide the required explanation, but it has the merit of stressing the need for establishing the correlation of chemical reactions that might represent the dynamic balance of exchanges in an organism that composes and decomposes, forms and destroys itself constantly. This opens the door to future theoretical models that will fulfill the objective of describing and explaining physiological metabolisms.

One of the aims of the analysis by which Senebier exploits experimental data is to establish that oxygen absorption is different and works out differently in warm-blooded and cold-blooded animals, but that all animals nevertheless, at various degrees, achieve a function of assimilation by their envelopes and muscles, independently from or in addition to the chemical reactions of pulmonary respiration, along with the latter's extension through blood circulation (Senebier 1807, II, 324–325). From this perspective, irritability, which constitutes a fundamental physiological property of animals, could not be attributed to a *vis insita* that would be a particular kind of vital principle, but "it [would] depend on the action and reaction of the humors that are constantly formed and excreted, as well as on the oxygen gas that muscles always absorb" (Senebier 1807, II, 326). It appears therefore essential to note that irritability and vital motions depend on such processes of oxygen fixation by organic fibers, and to study the variation of the corresponding functions in cases in which the oxygen supply is quantitatively reduced, and even suppressed, notably in cases of lethargy. Collecting experimental data relative to these chemical processes as they operate in organic structures fits an objective of unification in theoretical explanation. The specificity of this type of explanation in physiology, according to Senebier's pattern for it, is that it accounts for circular or alternating series of vital processes by subordinating them to a set of laws of the physical-chemical kind and to analytic models of organization underlining the functional complementarity of parts. Hence the orientation toward a specific theoretical interpretation:

> It is truly a beautiful sight that we are constantly offered by the uniformity of the laws to which all organized beings are subjected; everywhere one notes constancy and similarity in the relationships that link these beings with air, and although all these beings really differ in thousands of ways, air seems to always act on them in an equally indispensable and uniform manner for conservation of their life and destruction after their death: thus absorption of oxygen gas by all parts of animals and plants, discovered by Spallanzani, complements the indispensable necessity of this air and of oxygen gas for producing respiration and for generating animal heat and muscular irritability, since both subsist in animals deprived of their lungs; but we cannot observe without admiration how nature has disposed

> everything to achieve those grand aims, by giving muscles stronger affinities for oxygen gas than for blood which reddens them, so that blood does not seize the oxygen that comes to muscles by the skin or does only seize a small part of it; in this manner muscles rendered more irritable are more apt for motion, the circulation of blood is facilitated in the smaller vessels, and nutrition is favored by the heat produced which increases the force of affinity of organs for food molecules, etc.
>
> (Senebier 1807, II, 342–343)

The study of the phenomena of lethargy is called upon to corroborate this type of theoretical explanation, by taking into account processes of reduction, and even suspension, of normal functions. With the lowering of temperature, these phenomena result in a decrease and, later on, suspension of respiration, circulation, irritability, sensibility and nutrition, with conservation of a heat slightly higher than that of the milieu in which the lethargic animal is placed.[14] Once more, the general process is modulated into multiple variants according to the animals tested. In fact, following Spallanzani's practice, it is considered relevant to infer general effects from the observation of particular phenomenal sequences, provided these involve specimens belonging to different species, placed in circumstances in which environmental factors enable significant differentiations. Indeed, these modalities had been rigorously observed and noted by Spallanzani, but the question that still needed to be answered concerned an acceptable and valid theoretical explanation about the cause of lethargy. Changes of temperature due to intense cold seem to occasion most lethargic processes. These would be primordially due to diminution of muscular irritability that the cooling of tissues would provoke by hindering, even to the point of suspending, the gas exchanges metabolized by organic structures. The investigation of lethargic phenomena highlights at once the dominant themes about chemical processes alternatively synthesizing and decomposing organic structures, and about the functional integration of these processes according to modalities of organization that depend on the involved structures and operations.

Ultimately, in those studies of characteristic phenomena, Senebier tries concomitantly to systematize inferences drawn from the data gathered and to provide determining reasons for these in terms of physical-chemical processes specific to vital organizations. We thus perceive one of the keys, maybe the main one, to his way of interpreting the extension of the method of physiological analysis to a wide spectrum of living organisms; this key consists in framing up models conjoining the general physical-chemical relationships inferred from empirical induction, on the one hand, and conceptual schemes about the various modes of organization underpinning the required functional integration of processes, on the other hand (Senebier 1807, II, 389, 392).

In the antepenultimate section of another essay included in the *Rapports de l'air avec les Êtres organisés*, the *Mémoire de l'Éditeur relatif aux expériences de Spallanzani*, Senebier provides a remarkable overview on this composite rationality that characterizes explanations in general physiology. These explanations combine fundamental laws on chemical actions and reactions and records of difference in functional dispositions according to types of vital organization. Senebier supports a principle of uniformity in general relationships: "Organization yields in both realms the same effects: it holds the same relations with atmospheric air" (Senebier 1807, III, 344–345). Correlatively, the question arises about the possible variation of these relations depending on external and internal antecedent conditions and according to the modes of integration of processes that prevail among plants and animals. Analysis focuses then on similarities and disparities of processes between both. The final inference bred by this analytic approach is twofold: (1) The aimed-at theoretical explanation sets up in uniform order laws of organic functioning; and (2) it acknowledges on the other hand that types of organisms vary in accordance with the architectonic arrangement of their parts. It is presumed, though implicitly, that such complex arrangements result from standard combinations of their material components.[15]

Conclusion

By way of conclusion, I restate my objective, which was simply to show that Senebier may be attributed a role in announcing a general physiology that will later be located at the heart of biology, will develop stepwise during most of the nineteenth century and will be more precisely associated with research traditions stemming from Henri Ducrotay de Blainville, Johannes Peter Müller, Claude Bernard and Max Verworn, among others. Physiology, as conceived by Senebier, aimed at acceding to the rank of a general scientific discipline: (1) Based on observations and experiments, it would establish, by way of analysis, the general relationships, essentially physical and chemical, that rule over the elementary properties and operations of organized beings interacting with their external milieu; (2) it would from there ascend to law statements about the causation of functional processes in animals and plants; and (3) it would correlatively translate these general expressions of organic determinism according to the diverse and variable modes of integration of structures and functions among the several types of organisms. Physiology is thus made to appear to be a science about the elementary processes that characterize all life forms. To paraphrase a statement of Claude Bernard, it is not the least of Senebier's merits to have believed that "it is through the long sustained efforts of experimental analysis that we shall reach the elementary facts from which we will afterward deduce the synthetic conception that makes sense of the most diverse phenomena" (Bernard 1872, v).

Notes

1 On Senebier's methodology in his *Essai sur l'art d'observer et de faire des expériences*, see Duchesneau 2010, 2012, 590–598.
2 See Senebier 1800, III, 2: "L'anatomie des végétaux, en détaillant la structure des organes, conduit à la physiologie qui s'occupe de leurs rapports."
3 On the observation of chlorophyll by Senebier, see Canabal 2010.
4 On the physiology of respiration, see Holmes 1985; West 2015.

5 Senebier cites, in this instance, the memoir that Nicolas Théodore de Saussure had read at the Société d'histoire naturelle de Genève, in Year V, see Senebier 1800, III, 212–214.

6 In this instance, attempts at falsification are made use of to corroborate the explanation proposed.

7 See Senebier 1800, III, 264–265:

> Il serait très-difficile d'estimer la quantité de gaz oxygène rendu par les feuilles, parce que cela dépend d'une foule d'éléments qui sont très-variables: tels sont l'âge, l'état de la feuille, la saison, la sérénité du ciel, la quantité de l'acide carbonique dissous dans l'eau. J'ai donné dans mes divers ouvrages sur ce sujet, une foule d'expériences qui apprennent la quantité du gaz oxygène produit dans ces différentes circonstances & dans plusieurs autres que j'ai imaginées.

8 See Senebier 1800, III, 399:

> On comprend de cette manière comment la lumière retarde la germination; en décomposant l'acide carbonique, elle lui enlève alors l'oxygène qui ne se sépare qu'en très petite quantité, pendant que la plante est dans les ténèbres, mais qui favorise la fermentation en y restant; au lieu que lorsque la plante est au soleil, non-seulement elle la prive de cet oxygène, mais encore elle dépose dans les mailles de ses réseaux, une grande quantité de carbone qui est fortement antiseptique, qui donne au végétal de la rigidité en lui donnant plus de consistance, qui favorise le mouvement de ses fluides devenu nécessaire, & qui empêche une stagnation d'autant plus dangereuse qu'elle serait plus considérable dans une plante plus grande.

9 Respiration can in particular serve to illustrate this necessary distinction between vegetal and animal organisms: see Senebier 1800, V, 194.

10 See Senebier 1800, V, 231: "d'autant plus que les connaissances qu'on acquiert dans la physique & dans la chimie cultivées aujourd'hui avec tant de succès, sont des moyens puissants pour faciliter l'intelligence des phénomènes vitaux."

11 Concerning Haller's methodological pattern for physiology, see Monti 1990; Steinke 2005; Duchesneau 2012.

12 Concerning Senebier's personal relations with Bonnet, see Huta 1994; concerning Bonnet's theoretical views, see Duchesneau 2003 and 2006.

13 On Spallanzani's experimental methodology, see Monti 2005.

14 See in particular *Mémoire sur les animaux léthargiques*, §7, Senebier 1807, II, 368. Spallanzani's experiments on the physiology of respiration and his studies on lethargy were posthumously edited by Senebier and published in Spallanzani 1803. For an analysis of these, see Duchesneau 1982; for an analysis of Senebier's editorial work on Spallanzani's manuscripts on animal respiration, see Monti 2010.

15 See Senebier 1807, III, 346:

> On ne peut pourtant s'empêcher de remarquer cette uniformité de moyens pour conserver les êtres organisés; ils sont composés des mêmes éléments et ils ne paroissent différer que par leurs proportions dans les mélanges, c'est aussi pour cela qu'ils se servent réciproquement d'alimens; qu'ils se trouvent placés au milieu des mêmes substances et qu'ils sont toujours en rapports avec elles.

References

Bernard, C. 1872. *De la physiologie générale*, Paris: Hachette.

Canabal, M. 2010. Jean Senebier face à un organisme ambigu; recherches sur la matière verte. *Archives des Sciences, 63*: 47–54.

Duchesneau, F. 1982. Spallanzani et la théorie de la respiration: révision théorique. In G. Montalenti and P. Rossi (eds.), *Lazzaro Spallanzani e la Biologia del Settecento. Teorie, Esperimenti, Istituzioni Scientifice* (pp. 45–66), Florence: Leo S. Olschki.

Duchesneau, F. 2003. Charles Bonnet et le concept leibnizien d'organisme. *Medicina nei Secoli, 15*: 349–367.

Duchesneau, F. 2006. Charles Bonnet's neo-Leibnizian theory of organic bodies. In J. E. H. Smith (ed.), *The Problem of Animal Generation in Early Modern Philosophy* (pp. 285–314), Cambridge: Cambridge University Press.

Duchesneau, F. 2010. Senebier et le nouveau modèle d'analyse physiologique. *Archives des Sciences, 63*(3–4): 55–64.

Duchesneau, F. 2012. *La Physiologie des Lumières: empirisme, modèles et théorie*, Paris: Classiques Garnier.

Holmes, F. L. 1985. *Lavoisier and the Chemistry of Life: An Exploration of Scientific Creativity*, Madison, WI: University of Wisconsin Press.

Huta, C. 1994. Bonnet – Senebier: histoire d'une relation. In M. Buscaglia *et al.* (dir.), *Charles Bonnet, savant et philosophe (1720–1793)* (pp. 211–224), Geneva: Éditions Passé-Présent.

Ingenhousz, J. 1779. *Experiments upon Vegetables, Discovering Their Great Power of Purifying the Common Air in the Sunshine and of Injuring It in the Shade and at Night*, London: Elmsly and Pain.

Monti, M. T. 1990. *Congettura ed esperienza nella fisiologia di Haller*, Florence: Leo S. Olschki.

Monti, M. T. 2005. *Spallanzani e la rigenerazione animali: L'inchiesta, la communicazione, la rete*, Florence: Leo S. Olschki.

Monti, M. T. 2010. Senebier e i diari di Spallanzani sulla respirazione animale: un laboratorio di scrittura scientifica. *Archives des sciences, 63*(3–4): 113–128.

Senebier, J. 1782. *Mémoires physico-chimiques sur l'influence de la lumière solaire pour modifier les Êtres des trois règnes de la nature*, Geneva: Chirol.

Senebier, J. 1788. *Expériences sur l'action de la lumière solaire dans la végétation*, Geneva: Barde & Manget.

Senebier, J. 1800. *Physiologie végétale contenant une description des organes des plantes, & une exposition des phénomènes produits par leur organisation*, Geneva: J. J. Paschoud.

Senebier, J. 1802. *Essai sur l'art d'observer et de faire des expériences*, Geneva: J. J. Paschoud.

Senebier, J. 1807. *Rapports de l'air avec les Êtres organisés, ou Traités de l'action de l'action du poumon et de la peau des animaux sur l'air, comme de celle des plantes sur ce fluide ...*, Genève, J. J. Paschoud.

Spallanzani, L. 1787. *Œuvres*, Pavia.

Spallanzani, L. 1803. *Mémoires sur la respiration, traduits en français d'après un manuscrit par Jean Senebier*, Geneva: J. J. Paschoud.

Steinke, H. 2005. *Irritating Experiments: Haller's Concept and the European Controversy on Irritability and Sensibility, 1750–90*, Amsterdam, New York: Rodopi.

West, J. B. 2015. *Essays on the History of Respiration*, New York: Springer.

6 Organization and process. Living systems between inner and outer worlds

Cuvier, Hufeland, Cabanis

Tobias Cheung

Introduction

When Charles Bonnet described the monadic entity in his *Vue du Leibnitianisme* (1783) as a "perfect machine," in which "all pieces are necessary because they all act together for a common goal through mutual relations that they maintain between them and with the entire being … [as well as] with the surrounding monads and through them with the whole system,"[1] he referred to a specific agent model of systemic order that begun to play a central role in French and German comparative anatomy, physiology and medicine.[2] It was the model of a living thing as an individual agent that was able to maintain its inner order through its relations to the surrounding world. In this chapter, I would like to examine three variations of this model in the works of Georges Cuvier, Christoph Hufeland and Pierre-Jean-Georges Cabanis.

In the last decade of the eighteenth century, Cuvier worked as a comparative anatomist at the Jardin des Plantes in Paris, Cabanis was professor at the École de Médecine in the same city, and Hufeland taught medicine and physiology at the University of Jena.[3] Despite their differences – Cuvier's fixism, Hufeland's life force vitalism and Cabanis's focus on organic modifiability – they shared a common interest in the systemic order of living beings and their inside–outside relations. For Cuvier, Hufeland and Cabanis, the conditions of existence of organized bodies were based on the interrelation of the processual interaction of their parts with the world that surrounded them. In their agent models, living beings had to reproduce what they lost in the exercise of life, and their cycles of reproduction depended on a certain balance or harmony between their inner activity and the outer circumstances.

From Leibniz's monadic individualism to the materialism of the French idéologues, Cuvier's, Hufeland's and Cabanis's search for a new agent model was part of a *longue durée* transition of theo-ontologies into explanatory schemes of the immanent causation of agency through the inside–outside relations of self-active entities. Around 1800, multiple forms of agent models emerged not only in the works of their colleagues – for example in Xavier Bichat's, Jean-Baptiste Lamarck's and Johann Friedrich Blumenbach's physiologies – but also in anthropology, political economy, psychology and other domains of the emerging life sciences.[4]

Cuvier, Hufeland and Cabanis did not develop a detailed theory of the milieu that relied on the systemic unity of outer circumstances, but they wanted to know why and how their agents interacted with the world that surrounded them.[5] They thus asked for the conditions and for the mode of existence of living beings – or organisms, as they were often called at the end of the eighteenth century.[6] Within this discursive framework, Cuvier focussed on the inside–outside relations of systemic orders of correlated and subordinated parts, Hufeland on reproductive processes and cycles of destruction and production that depended on the exchange of inner and outer particles, and Cabanis on complex networks of organic units with specific stimuli–reaction schemes that mediated between inner and outer worlds.

Cuvier's science of organized beings

Like his colleague Vicq d'Azyr, Cuvier sought to transform Buffon's natural history into a new empirical science based on the works of comparative anatomists and physiologists.[7] In *Tableau Élémentaire* (1798) and *Leçons d'anatomie comparée* (1800–1805), Cuvier combined the processual order of "composition and decomposition" of circulating fluids with the mechanical order of bones, tendons and muscles.[8] In his "science of organized bodies," the physiological and anatomical properties of living beings constituted a single "system" of organs that carried out acts in order to sustain the activity and the "organization" of the whole body. With respect to their role in organized systems, Cuvier called such acts "functions." As "carriers" of "functions," organs possessed specific physical structures, and they systematically interacted with all the other organs.[9]

Within each organized being, the systemic unity of organs formed a "sort of a circle."[10] Each organ sustained the operations of the other organs, which supported again its own activity.[11] In Cuvier's circle, five organic ensembles constituted in general the bodies of higher animals: the digestive, respiratory, sensitive, locomotive and circulatory apparatus.[12] The variety of their different operations formed a "series of decompositions and reconstructions."[13] This internal series had to correspond to an outer referential system, beyond the surface of the body, because the maintenance of the "inner condition" of coordinated "acts" was for Cuvier only possible when the living body could regain "from the bodies which surround it what it lost in the exercise of life."[14] The loss was continuous. Each life-maintaining act effected a change of the involved "carriers." This change could refer to the "energy" of an organ or to mechanical deteriorations of organic parts used within the body or the surrounding world.[15] According to Cuvier, the most "important phenomenon of life" was the activity of an internal "stove" that represented the "continuous circulation from the outside to the inside, and from the inside to the outside."[16]

In Cuvier's science of organization, the existence of living beings depended on two factors: on the processual order of their inner organic apparatus and on their inside–outside relations. The twofold referential system of "inner" and "outer conditions of existence" (*conditions d'existence intérieures, extérieures*) constituted for Cuvier the "rational principle" of natural history and thus of his new science:

> Natural history has … a proper rational principle which is peculiar to it, and which it applies advantageously on many occasions; it is one of the conditions of existence, commonly called final causes. For the same reason that nothing can exist if it does not provide the conditions which render its existence possible, the different parts of each living being have to be coordinated with each other so as to render the entire being possible, and this not only in itself, but also in its relations with those beings which surround it. The analysis of these conditions often leads to general laws as demonstrable as those which are derived from calculation or experiment.
>
> (Cuvier 1817, vol. 1, 6)[17]

From Cuvier's perspective, naturalists had to consider the order of organized bodies not only within the borders of their individual unity but also with respect to their relation to the "surrounding bodies" (*corps environnants*).[18] He combined the manifold relations of the seemingly "infinite number of fluids, forms, characters and dispositions"[19] within animals with their "way of life" (*genre, manière de vie*) in the outer world. The relations between the different acts of their "way of life" corresponded again to the "whole organization" of their inner order. Without such a correspondence, organized bodies could not maintain the processual framework that kept them alive: "The whole organization of the animal is in a necessary harmony to its way of life. The organic apparatus of its jaws has to correspond with its food, and thus with its whole organization" (letter from Cuvier to Hartmann on May 18, 1791, in Duvernoy 1833, 125). Systemic differences and modifications of the inner organization of animals necessitated "an equal one in the way of life."[20] As a comparative anatomist, Cuvier dissected the bodies of dead animals, but he had to be able to recognize in the dead what was alive, and to read the "conditions of life" in the inanimate tissues of lanced bodies:

> The general form of the feet of insects depends on their way of life. Are they destined to remain in the water, to swim? Then they are flat, long and hairy. Are they made to dig in the earth? Then they are bigger, serrated and cutting. Serve they only to move on the earth? Then they are long and cylindrical.… After judging all this, one can precisely recognize, even in the dead insect, its behaviour and its way of life.
>
> (Cuvier 1800–1805, vol. 3, 452–453)

For Cuvier, the "way of life" of each living being was determined through its "type of organization." He constructed the anatomical order of these types according to two "laws of coexistence": the "principles of subordination" and of "correlation." The principle of subordination relied on the "influences" of certain organs on the activity of other organic parts. In higher animals, the brain and the heart had, for example, a stronger regulatory "influence" on the overall activity of organs than the ear or the gall bladder.[21] According to the second principle, the "combinations" of the "main organs" (*organes principaux*) of types were "correlated" in such a way that they could systematically interact.

In Cuvier's animals, the subordination and correlation of organic parts could not be changed without destroying the entire system. On the basis of such systemic orders or "plans," each animal represented a specific "type of organization":

> Every organized being forms a whole, a unique and closed system, in which all the parts correspond mutually, and contribute to the same definitive action by a reciprocal reaction. None of its parts can change without the others changing too; and consequently each of them, taken separately, indicates and gives all the others.
>
> (Cuvier 1798, 16)

The principles of subordination and of coordination established for Cuvier not only the "inner conditions of existence" of living beings but also the limits of the "variations" of "types of organization." In his view, secondary organic parts might vary "infinitely" in their forms, but these "variations" could not transcend the "laws of coexistence" between the "main organs."[22] Within this universal harmony, the order of each "type of organization" was determined "from its origins," and it lasted until the "actual order" of nature was destroyed.[23]

As a gross anatomist, Cuvier was mainly interested in the comparison of the systemic orders of "types of organizations" of higher animals and in the limits of their modifiability. When he discussed the foundations of his "science of organized bodies," he often referred to, but rarely examined in detail, the processual framework of the cycles of production and reproduction between the inner activity of organs and the different transformations of anorganic and organic materials. Within the context of emerging physiologies around 1800, the dynamic order of his organic "stove" played again a central role in Hufeland's medical theory of reproductive organisms.

Hufeland's reproductive organisms

In his *System der practischen Heilkunde* (1800–1805), Hufeland focussed on "irritable living beings."[24] He called such beings "organisms." With reference to Newton's gravitational force, Hufeland thought – like his colleague Johann Friedrich Blumenbach – that these bodies possessed a specific life force, itself nonreducible to other physical forces, that would explain their order-generating processes.[25] According to Hufeland, the dynamics of each life force that "expressed" itself as a regulating entity within individual organic bodies depended on their "inner state of organization" and on the order of things in the "outer world" (*Außenwelt*). The inner organization and the outer world thus represented the "inner" and "outer conditions of life."[26]

Against John Brown's thesis in the *Elementa medicinae* (1780), that all inner states of life were forced states determined by outer stimuli, Hufeland argued that the "self-activity" of organisms was determined not only by outer circumstances but also by the inner "activities" of their life forces.[27] In his view, organisms as self-active beings could both produce stimuli within themselves and react to stimuli from the outer world:

> The organism is not just a passive being, determinable and determined by outer things, but something that is self-active, determined by itself, thus a being that is itself active when it is affected by the outer world, that reacts upon the outer things and that modifies their effects in various ways.
>
> (Hufeland 1811, 18)[28]

As a regulating entity within organisms, Hufeland's life force "consumed" through its activity a kind of energy that was present in the physical organization of organic parts and in their chemical reactivity. The consumption resulted into a partial "destruction" and "dissolution" of these energetic "dispositions." According to Hufeland, organisms died when a certain "disorganization" prevented the "reproduction" of these dispositions.[29] Different "states of life" resulted from the manifold constellations and interactions between inner organizations, life forces and the influx of particles from the outer world. If these states changed the "normal" activity of a life force, Hufeland called them "pathological" or "diseased states." The operational mode of Cuvier's stove thus became a central issue in Hufeland's medical physiology and pathology.[30]

Within Hufeland's living bodies, the "continuous excitation or activation of the life force" maintained a cyclic process of consumption and reproduction of the material disposition that made the expressions of the force possible.[31] The production of inner stimuli by the force as well as its reactions caused "dissolutions" and consumed energy, yet these stimuli and reactions could initiate processes that reproduced its energetic dispositions.[32] Hufeland therefore combined the processes of "excitation," "self-consumption" and "self-restauration" (or "self-production") into a single self-maintaining process that depended on the "transformation" of "foreign matter" for the production and the reproduction of the physical basis of the expressions of life forces.[33]

As self-active, self-stimulating and self-reproductive agents, Hufeland's organisms were able to regulate the permanent "influx of outer material substance" and the "continuous change" of their components: "The components of our bodies are constantly replaced, they part from us by excretion, and they return through the air and the nutrition. It is the process of life itself that necessitates a continuous change of these components" (Hufeland 1795, 65). The systemic connection between the different processes of excitation, consumption and reproduction constituted for Hufeland the material basis of "all phenomena of life."[34] It represented the embodied "self" of an organism, and this self had to constantly renew itself. The self-activity of the organism was both the "product" and the "maintaining cause" of its "organization":

> The process of life is as a whole and in each moment basically composed out of two contrasting operations, self-consumption and self-production, which both constitute life, and yet they are only possible through it, regulated and determined through its expressions (excitation).
>
> (Hufeland 1800–1805, vol. 1, 3)[35]

Based on the passage of stimuli and substances between the outer world and the inner organization, the relation between Hufeland's "outer" and "inner conditions of life" was not a simple representation of the outer order of things within the organism. Rather, their combinations and properties underwent a radical change. They were actively "assimilated" to its inner order. According to Hufeland, chemical analyses could explain the different material components of life, but not the "living being itself" as an "assimilating" and "synthesizing" agent.[36] In the inner world of the organism, "all kinds of substances" were put into a "very particular relation" to each other, a relation that did not exist in "inanimate nature."[37] The regulating activities of life forces "modified" and "annulated" the chemical "bondings" of nonorganic molecules.[38] Through this "transformation," the entire organic body could physically resist the "dissolving chemical forces" and "reproduce" itself.[39]

The relations between various forms of mechanical and chemical "transformations" – such as "assimilation," "digestion," "secretion" and "excretion" – constituted a "system" of interrelated and interacting reproduction processes. The organism thus "created" itself as an organized and processual, individual body. Its body performed acts in the outer world, yet these acts were always part of an "inner organic circle that returned to itself" (*in sich selbst zurückkehrender organischer Zirkel*).[40]

The reproduction processes of Hufeland's organism maintained not only a permanent flux and transformation of particles but also a certain "state of equilibrium" between its physical organization and its inner activities. This "inner equilibrium" had again to correspond to the "outer conditions of life."[41] The "inner" and "outer conditions" of the life of an organism were thus also the conditions of the reproducibility of a double equilibrium. Like Cuvier, Hufeland thought that the outer conditions could modify but not substantially change the order of "inner organizations" of organic bodies without destroying their lives.[42]

As a medical doctor, Hufeland was not primarily interested in debates about the limits of possible modifications of types of organization of living beings. Rather, he examined their manifold inner "states of life" and the interrelation of these states with different "modes of life" as habitualized patterns of activities in the outer world. For Hufeland, the "normal state of life" as a "state of health" correlated with the natural "state of equilibrium" prescribed by the organization of each organism.[43] Illness was thus an "aberration" of the "normal state of life," and healing a process of "recovery."[44] In his macrobiotics, he discussed the impact of various "modes of life" that had been modified according to the six factors of Galen's *regimen sanitatis* – light and air, food and beverage, exercise and rest, sleeping and wakening, secretion and excretion and affections of the soul – on the inner "state of health" of humans in order to prolong their life through an optimal balance of inside–outside movements.[45]

Like Hufeland, Cabanis discussed the specific connections between excitability and reproduction that maintained cycles of consumption and regeneration in living beings. However, his main focus was to explain the systematic order of organic bodies through a complex network of multiple "reactive centers" (*centres de réaction*), in which circulated different kinds of stimuli and sensations.[46] Futher on, Cabanis integrated the agent model of living beings into a physiological anthropology, in which the special role of humans among other animals depended on the transformation of systemic inside–outside relations into reflexive faculties.

Cabanis's organic networks of reaction centers

In *Rapports du physique et du moral de l'homme* (1802),[47] Cabanis focussed on the way different reaction centers within organic bodies produced, influenced and reproduced the activity of entire "living systems." His reaction centers had in general three basic properties: they could receive impressions, transform them through "inner acts" (*actions intérieures*) into their own inner order, and react to them through "outer acts" (*actions extérieures*) in the "worlds" that surrounded them. These "worlds" represented both the "networks" (*réseau*) of other interacting reaction centers, such as organs or organic subsystems, or the order of things outside of the living body itself.

In human bodies, reaction centers – for example the brain or the stomach – possessed different sensibilities, exhibited different reactions and produced different products (sentiments, nutrition, thoughts). Furthermore, the organization and capacities of reaction centers were different in humans and higher animals. However, all reaction centers operated according to the same processual framework of "inner" and "outer acts." It was thus the same logic that defined humans and animals as living systems and that distinguished humans from animals through specific properties.[48]

In *Recherches physiologiques sur la vie et la mort* (1800), Xavier Bichat had drawn a distinction between the "animal life" and the "organic life" of humans. According to Bichat, their "animal life" relied on acts in the outer world that were regulated by the nervous system of the brain, while their "organic life" relied on the activity of the inner organs and the spinal cord.[49] However, he also indicated that this was merely an "idealized" distinction. Within living beings, both lives interacted.[50] In the *Rapports*, Cabanis examined precisely the order of these interactions. Like Paul Thiry d'Holbach in his *Système de la nature* (1770), he sought to explain how the "physiological" and "intellectual operations" of human beings interacted as physical orders of a single system.[51]

Rejecting Haller's dualism of the sensibility of nerves and the irritability of muscles, Cabanis adopted for the operational mode of his reaction centers the model of Théophile de Bordeu, who held that each organ in a living system had its own "sensibility."[52] While Cabanis considered the nervous system the "specific seat" of sensibility in higher animals, many lower animals, such as polyps, could nonetheless produce movements through stimulus–response mechanisms without such a system.[53] In order to explain both phenomena, he referred to a form of reactive "energy" that, like electricity, is mediated through substances without moving them.[54] If such an "energy" could remain in amputated limbs, it would produce movements until it is "consumed." Irritability was thus merely a consequence of sensibility.[55]

Cabanis held that human sensibility produced not only movements but also different "kinds of impressions" or "sentiments." These sentiments were again modified within the body. Sentiments that entered the brain through the nervous system triggered "intellectual operations." These operations in turn could transform sentiments into "perceived impressions" (*impressions perçues*).[56] However, not all sentiments were directed toward the brain. Many of them oscillated between the inner organs.[57] Cabanis described these sentiments as "unperceived impressions" (*impressions inaperçues*).[58] In the processual framework of "intellectual operations," "inner sentiments" just represented "confused impressions" (*impressions confuses*). As "unperceived impressions," however, such "confused impressions" had a precise role within the "play" of inner organs. They constituted the animality of humans.

In Cabanis's view, the production of ideas and the operations of the intellect relied on the transformation of sentiments. However, Cabanis criticized Locke and Condillac for focussing exclusively on the "distinct sentiments" (*sentiments distincts*) of sensory organs and not on the interaction of different kinds of sentiments in living bodies.[59] According to Cabanis, a good philosopher should "classify" sentiments "in such a way that one could assign to each organ the inner impressions that are proper to it."[60] Using this classification, the philosopher could then determine the different roles of sentiments in a living system.

Cabanis called the outer sentiments of the sensory organs "sensations" (*sensations*), which were transported to the brain through specific nerves. On their way to the brain, however, they might come into contact with "inner sentiments" circulating in the body. In general, outer sentiments reached the brain more directly than inner sentiments, but they were not transported like messages through a pneumatic tube system. Rather, they often interacted with various reaction centers, such as ganglia, that have a life of their own and might function as mediating nods for both kinds of impressions. When sensations finally arrived in the *sensorium commune* of the brain, they could be "mixed" with other sentiments. Strong inner sentiments, such as instinctive needs or passions, might moreover modify "intellectual operations" and "voluntary acts." Finally, very weak inner sentiments, which passed into the brain "unnoticed" by the intellect or will, could also influence their operations. Cerebral "operations" thus depended in various ways on the "play of organs" and their "sensibility without sensation": "In this limited sense of the word *sensation*, it is beyond doubt that only some of our ideas and resolutions come from sensations; many of them are due to inner impressions resulting from the play of different organs" (Cabanis 1823–1825, vol. 3, 449–450).[61]

With his critique of Locke and Condillac, Cabanis defined his objective in the *Rapports*: he intended to examine the "production," "cooperation," "enchainment" and "subordination" of different kinds of sentiments and movements across a range of "physical operations" that covered both "intellectual" and "physiological operations." All these operations were executed by "inner" and "outer acts" of reaction centers, and these centers operated like a single "network" of multiple interacting and interrelated entities in living systems. Living systems in turn interacted as "total systems" (*systèmes totaux*)[62] with the outer world.

In these systems, each reaction center functioned as a "central point" that established a "sphere of activity" (*sphère d'activité*) within which it reacted to stimuli and regulated movements.[63] Between stimulation and reaction, however, stimuli were "transformed" by an assimilative process into proper entities of the center. Cabanis called this process "metamorphosis."[64]

While each center had its own particular sensibility and organization, the general framework of their activity was always the same: stimulation, transformation and reaction. Cabanis could thus compare the brain, which transformed impressions into perceptions, to the stomach, which transformed food into digestible material.[65] Besides the brain or the "thinking organ," Cabanis cited three other "main centers" in the body of human animals: the "phrenic region" of the diaphragm and the stomach, the "hypochondric region" of the liver, spleen and upper intestines, and the "region" of the reproductive organs, the urinary system and the lower intestines.[66] These three "regions" exerted a "considerable influence" on the "intellectual operations" of the brain. However, they also interacted within a single "total system." Through this "order of correspondence" (*ordre de correspondance*), the centers maintained and reproduced the activities that continuously decomposed and recomposed organic material.[67]

Like Erasmus Darwin in *Zoonomia* (1794–1796) and Jean-Baptiste Lamarck in *Philosophie zoologique* (1809), Cabanis was also interested in certain modes of interactions with the outer world that could result in changes to the inner organization.[68] These changes were part of their "physiological histories" (*histoires physiologiques*).[69] Cabanis described the ensemble of interactions of human animals with the outer world as their "way of life" (*genre de vie*). This depended on three factors: first, on the general inner disposition of outer acts, also called "temperament"; second, on "particular" or "accidental circumstances" that were part of the "ensemble of physical circumstances" (*ensemble des circonstances physiques*) that constituted the outer world; and, third, on living entities as agents that made use of the "circumstances."[70] For these agents, the "randomness of circumstances" (*hasard des circonstances*)[71] represented potential "chances in life" (*chances de la vie*),[72] that is, opportunities to act in order to maintain their inner order, although certain "circumstances" might also prevent such acts. The outer world was thus not only an "ensemble of physical circumstances," but also an ensemble of "circumstances of life" (*circonstances de la vie*).[73]

The choices that human animals made within the spectrum of changing outer circumstances could result in "physical habits" that differed from the "habits" of their "primitive temperaments." According to Cabanis, "habits" were "ensembles" of coordinated "acts" in the outer world. Like Hufeland, he called them also "regimens" (*régimes*) or "plans of life" (*plans de vie*).[74] They were shaped by a variety of inner and outer factors, such as instincts, natural and "social" circumstances, and "conscious" choices. If the continuous use of these factors over a long period resulted in new "habits," these "habits" might not only produce "accidental dispositions" (*dispositions accidentelles*)[75] in some organs but also modify the "general plan of organization":

> We have thus recognized that the general expression *regimen* includes all the physical habits taken together; and we know, moreover, that these habits are capable of modifying and even of changing not only the organs' mode of action but also their inner states and the character of the inclinations of the living system. In fact, it is well known that the plan of life can, according to whether it is good or bad, considerably improve the physical constitution, or can alter and even destroy it without recourse. Through this influence, each organ can fortify or weaken itself; its habits can be perfected or dissipated from day to day. The impressions through which the order of the stabilizing movements is reproduced, impressions that incessantly tend to introduce new series of movements, are themselves capable of undergoing notable changes. Just as, through the advantageous or harmful effect of the regimen, the organs take on new ways of being and of acting, they also take on new ways of feeling. Finally, even when the initial change has been circumscribed and local, these modifications of sensibility are most often reproduced, as it were, by the entire living system.
>
> (Cabanis 1823–1825, vol. 3, 12–13)[76]

Habitualized outer interactions, which modified the "total system" and the "primitive disposition of sensibility," also created new inner temperaments. Such "acquired temperaments" (*tempéraments acquis*) could be "transmitted" to the next generation.[77] Cabanis thus entered into the speculative realm of the "transformation" of "primitive organizations." However, like Étienne de Lacépède in his *Discours sur la durée des espèces* (1800), Cabanis supported a Cuverian taxonomy of natural types (based on the resistance of "primitive organizations" to modifying outer influences), although he did not neglect the fact that accumulated "accidental changes" could result in "important transformations" (*transformations importantes*).[78]

Concluding remarks

In their agent models of living beings, Cuvier, Hufeland and Cabanis focussed on systemic orders and inside–outside relations. They distinguished between the inner order of organic beings and the world that surrounded them. As inner and outer conditions of existence, both orders were systematically interrelated and at the same time separated from each other through the borders of a body that maintained its existence through cyclic processes of production and destruction.

These border-transcending processes depended on an influx of particles from the outer world into an organized entity of solids and fluids.

Cuvier was mainly interested in the anatomical organization of organs, muscles and bones according to principles of correlation and subordination. However, he was convinced that the most important phenomenon of life was the activity of an internal "stove" that continuously moved and replaced parts from the outside to the inside, and from the inside to the outside. Hufeland and Cabanis referred to the physiological activity of this stove, but not as comparative anatomists that sought to define the possibility of the existence of living beings as closed systems of interrelated parts. Rather, they looked at organic systems in order to explain their multiple inner states and the influence of their different ways of life on these states – without neglecting that this modifiability of their inner organization occurred within the limits prescribed by Cuvier's logic of only two states: dead or alive. While Hufeland explained the change of inner states through the different modes of activity of a life force that expressed its regulative potential in the dynamics of reproductive inside–outside processes, Cabanis examined complex organic networks of multiple interactive centers with specific stimuli–reaction schemes.

Cuvier's, Hufeland's and Cabanis's animals were self-active, self-organizing and self-stimulating agents. The self of their agents represented the systemic order of cyclic processes between inner and outer worlds operated by various regulating entities with proper energies or sensibilities. As creative and self-transforming agents, they produced, within the limits of their conditions of existence, their own histories of inside–outside relations in reproducing themselves through an ongoing series of acts according to the changing circumstances of outer worlds.

All their outer acts were part, as Hufeland had put it, of an "inner organic circle that returned to itself," and yet these same acts consumed what the inner organic circle was maintaining. Living bodies had to reproduce, in Cuvier's words, "what they lost in the exercise of life." The rational order of life thus became a processual and organized order of cycles of production and dissolution, and these cycles involved inside–outside relations of continuous transformations – such as assimilation, digestion, secretion and excretion.

Cuvier's, Hufeland's and Cabanis's living beings thus were self-active agents that interacted with the outer world. Without these interactions, they would die, and through them they had their own ways of life. Within this existential constellation, their acts had always a double meaning. They were part of the dynamics of the inner order of interacting organic parts – the networks of Cabanis' reaction centers – and of the manifold interactions of the entire body with the world that surrounded it. These agent models of inside–outside relations and the different aspects of the systemic order of their conditions and modes of existence played not only a central role in various emerging biological domains – such as experimental physiologies, cell theories, evolutionary theories and ecologies – but also, besides the philosophies of Kant, Schelling and Hegel, within the French context of the first half of the nineteenth century in Destutt de Tracy's, Saint-Simon's and Comte's anthropologies and sociologies.[79]

Notes

1 Bonnet, *Vue du Leibnitianisme*, in Bonnet 1779–1783, vol. 8, 306.
2 Cf. Anderson 1982; and Cheung 2004 and 2006a.
3 For further bio-bibliographical references, see Szyfman 1982; Outram 1984; Corsi 1988; Schönfeld 1988; Pfeifer 2000; and Taquet 2006.
4 Cf. Manuel 1956; Figlio 1976; Düsing 1986; Warnke 1998; Wolfe 1999; Blanckaert 2006; Cheung 2008 and 2014; and Hanulak 2009.
5 Canguilhem 1998, 129–154 reconstructed the history of milieu-theories in the French context. For the German context, see Trepl 1994.
6 For the usage of the expression and different concepts of "organism" in the eighteenth century, see Robin 1880; Köchy 1995; Kucharczik 1998; Cheung 2000, 2006b and 2010a; and Wolfe 2010.
7 Cf. Vicq d'Azyr 1805, vol. 4, 35:

> L'enseignement de l'Anatomie peut être séparé de celui de la Physiologie, comme, en physique, on peut examiner les differentes parties d'une machine, sans rechercher quels en sont les usages. Mais enseigner la Physiologie sans l'Anatomie, ce seroit s'éloigner des connoissances qui peuvent seules être les bases d'une saine théorie; ce seroit ouvrir de toutes parts un champ libre à l'erreur.

For the French context of natural history, comparative anatomy and physiology, see Lepenies 1976 and Schmitt 2009.
8 For biographical references and detailed analyses of Cuvier's work, see Coleman 1964; Outram 1984 and 1984; Cheung 1999; and Taquet 2006.
9 Cf. Cuvier 1800–1805, vol. 1, 33; and ibid., vol. 3, 10.
10 Ibid., vol. 1, 46.
11 Ibid., vol. 3, 4.
12 Ibid., vol. 1, 54–55.
13 Ibid., vol. 5, 1.
14 Ibid., vol. 3, 2.
15 Cf. ibid., vol. 1, 35 and 46.
16 Ibid., vol. 1, 4–5.
17 Unless otherwise indicated, all translations in this chapter are the author's own.
18 Ibid.
19 Cuvier 1798, 6.
20 Cf. Cuvier 1800–1805, vol. 3, 460:

> lorsque la longueur du canal intestinal s'écarte beaucoup dans un animal de celle observée dans les animaux voisins, dont le genre de vie est à peu près le même, le diamètre de ce même canal augmente ou diminue souvent d'une manière inverse, et détruit, en partie, l'effet d'une semblable diminution ou augmentation dans la longueur; sinon le genre de vie de l'animal en est modifié.

21 Cf. Cuvier 1798, 16.
22 Cf. Cuvier 1800–1805, vol. 1, 58:

> Au reste, en demeurant toujours dans les bornes que les conditions nécessaires de l'existence préscrivoient, la nature s'est abandonnée à toute sa fécondité dans ce que ces conditions ne limitoient pas; et sans sortir jamais du petit nombre des combinaisons possibles entre les modifications essentielles des organes importans, elle semble s'être jouée à l'infini dans toutes les parties accessoires.

23 Cuvier 1812, 58. For Cuvier's role in the French debates about the transformation of species, see Appel 1987 and Rudwick 1997.
24 For biographical references and detailed analyses of Hufeland's work, see Pfeifer 1968 and 2000; Schönfeld 1988; Mayer 1993; and Wiesing 1995, 66–81.

25 For the German debate about life forces, Blumenbach's physiology and Newtonianism in the eighteenth century, see Blumenbach 1786; Lenoir 1980 and 1982; Reill 2005; and Wolfe 2014.
26 Cf. Hufeland 1795, 15; and Hufeland 1800–1805, vol. 1, 51.
27 For a comparison between Brown's and Hufeland's medical theories, see Risse 1971; Henkelmann 1981; and Tsouyopoulos 1988.
28 Cf. Hufeland 1800–1805, vol. 1, 71.
29 Hufeland 1795, 53. cf. ibid., 71–72; and Hufeland 1800–1805, vol. 1, 60.
30 For the German and French context of Hufeland's medical theory and pathology, see Pfeifer 2000.
31 Hufeland 1800–1805, vol. 1, 2, 143. cf. Hufeland 1860, 41:

> Jedes Leben ist folglich eine fortdauernde Operation von Kraftäußerungen und organischen Anstrengungen. Dieser Prozeß hat also nothwendig eine beständige Consumtion oder Aufreibung der Kraft und der Organe zur unmittelbaren Folge, und diese erfordert wieder eine beständige Ersetzung beyder, wenn das Leben fortdauern soll. Man kann also den Prozeß des Lebens als einen beständigen Consumtionsprozeß ansehen, und sein Wesentliches in einer beständigen Aufzehrung und Wiederersetzung unserer selbst bestimmen.

32 Cf. Hufeland 1800–1805, vol. 1, 234.
33 Cf. Hufeland 1799, 54. cf. Hufeland 1800–1805, vol. 1, 215–216:

> Erregung also bewirkt den innern Lebensprozess, dieser die organische Mischung und Darstellung der Stoffe, und diese producirt wieder die Erregbarkeit, ohne welche keine Erregung möglich wäre. Folglich ohne Reiz keine Erregung, ohne Erregung kein Organismus und keine Erregbarkeit, aber ohne Erregbarkeit auch keine Erregung, und eben so wenig würde beydes ohne das Geben und in einander Wirken solcher Stoffe möglich seyn, die zum Material des Organismus geschickt sind.

34 Hufeland 1811, 9–10.
35 Cf. ibid., 61–64 and 215–216.
36 Cf. Hufeland, *Atmosphärische Krankheiten* (1824), in Hufeland 1834, 305.
37 Hufeland 1800–1805, vol. 1, 59–60.
38 Hufeland 1795, 63–64; and *Atmosphärische Krankheiten*, in Hufeland 1834, 304–305.
39 Cf. Hufeland 1860, 31–32.
40 Hufeland 1800–1805, vol. 1, Preface, xi–xii; and *Atmosphärische Krankheiten*, in Hufeland 1834, 305.
41 Hufeland 1799, 83–84. Cf. Hufeland 1800–1805, vol. 1, 51.
42 Cf. ibid.; and Hufeland, *Die Welt des Lebens* (1815), in Hufeland 1834, 54.
43 Hufeland 1795, 2–3.
44 Hufeland 1851, 43.
45 Hufeland 1795a, 138; and 1860, 42–44. For Hufeland's macrobiotics and its Galenic context, see Schipperges 1987.
46 For the wider context of the French and German debates about stimuli–reaction schemes and sensations, see Möller 1975; and Starobinski 1999.
47 For biographical references and detailed analyses of Cabanis's work, see Staum 1980; Besançon 1997; and Cipollini 1998.
48 For the relationship between medicine, physiology and anthropology around 1800, see Figlio 1975; and Blanckaert 2006.
49 Bichat 1801, vol. 1, 115–116. Cf. Bichat 1802, 2–4.
50 Ibid.
51 Cf. D'Holbach 1770, vol. 1, 2:

> On a visiblement abusé de la distinction que l'on a fait si souvent de l'homme *physique* et de l'homme *moral*. L'homme est un être purement physique; l'homme moral n'est que cet être physique considéré sous un certain point de vue, c'est-à-dire, relativement a quelques-unes de ses façons d'agir dues à son organisation particulière.

For the relationship between Cabanis and the French Idéologues, see Moravia 1966 and 1974; Staum 1974; Yolton 1991; and Guichet 2010.

52 In his essay on the guillotine (1795), Cabanis still relied on Haller's distinction between irritability and sensibility. cf. Cabanis 1795; and Chazaud 1998. For the different sensibility-debates, see Steinke 2005; Cheung 2010b; Gissis 2010; and Wolfe 2014.

53 Cf. Cabanis 1823–1825, vol. 4, 272:

> Les recherches les plus attentives de l'anatomie moderne n'ont pu faire découvrir de nerfs ni d'appareil cérébral dans quelques animaux imparfaits, tels que les polypes et les insectes infusoires: cependant, ces animaux sentent et vivent; ils reçoivent des impressions qui déterminent en eux une suite analogue et régulière de mouvements.

54 For the role of animal electricity in models of organic order in the eighteenth century, see Bernardi 1992.

55 Cf. Cabanis 1823–1825, vol. 3, 115.

56 Ibid., vol. 4, 303.

57 Ibid., 274.

58 Ibid., 408.

59 Ibid., 311–314 and 465.

60 Ibid., 465.

61 Cf. ibid., vol. 4, 225–226 and 273–276.

62 Ibid., 305.

63 Cf. ibid., 399:

> Les organes ne sont susceptibles d'entrer en action, et d'exécuter certains mouvements, qu'en tant qu'ils sont doués de vie, ou sensibles: c'est la sensibilité qui les anime; c'est en vertu de ses lois qu'ils reçoivent des impressions, et qu'ils sont déterminés à se mouvoir. Les impressions reçues par leurs extrémités sentantes sont transmises au centre de réaction; et ce centre, partiel ou général, renvoie à l'organe qui lui correspond les déterminations dont l'ensemble constitue les fonctions propres de cet organe.

64 Ibid., vol. 3, 160–161.

65 Cf. ibid., 180–181; and ibid., vol. 4, 318 and 324.

66 Ibid., 446.

67 Ibid., 15.

68 For the relations between Erasmus Darwin, Lamarck and Cabanis, see Moravia 1974, 63–74; and Szyfman 1982.

69 Cabanis 1823–1825, vol. 3, 101.

70 Ibid., 376; and ibid., vol. 4, 140–141.

71 Ibid., 248.

72 Ibid., vol. 3, 8.

73 Ibid., vol. 4, 161–162.

74 Ibid., 12–13.

75 Ibid., 430.

76 Translated by George Mora in Cabanis 1981, vol. 2, 656, modified by T. Cheung.

77 Cf. Cabanis 1823–1825, vol. 4, 146–147:

> Mais l'empire des habitudes ne se borne pas à ces profondes et ineffaçables empreintes qu'elles laissent chez chaque individu, elles sont encore, du moins en

partie, susceptibles d'être transmises par la voie de la génération. Une plus grande aptitude à mettre en jeu certains organes, à leur faire produire certains mouvements, à exécuter certaines fonctions; en un mot, des facultés particulières, développées à un plus haut degré, peuvent se propager de race en race.

78 Ibid., 6–8. cf. ibid., 161–162 and 248–250. For the influence of Lacépède on Cabanis, see Staum 1980, 187–188.

79 Cf. Greene 1969; Haines 1978; Lesch 1984; Duchesneau 1987; Zammito 1992; Weingarten 1997; and Guillo 2003.

References

Anderson, L. 1982. *Charles Bonnet and the Order of the Known.* Dordrecht, Boston, MA, London: D. Reidel.

Appel, T. A. 1987. *The Cuvier-Geoffroy Debate. French Biology in the Decades before Darwin.* New York, Oxford: Oxford University Press.

Bernardi, W. 1992. *I fluidi della vita. Alle origini della controversia sull'elettricità animale.* Florence: L. S. Olschki.

Bichat, X. 1801. *Anatomie générale appliquée à la physiologie et à la médecine.* 4 vols. Paris: Brosson & Sarrazin.

Bichat, X. 1802. *Recherches physiologiques sur la vie et la mort.* Paris: Brosson & Gabon.

Blanckaert, C. 2006. Le "circuit" de l'anthropologie. Figures de l'homme naturel et social dans le système méthodique des savoirs (1782–1832). In C. Blanckaert and M. Porret (eds.), *L'encyclopédie méthodique (1782–1832). Des lumières au positivisme* (pp. 69–102). Geneva: Droz.

Blumenbach, J. F. 1786. *Institutiones physiologicae.* Göttingen: Johann Christian Dieterich.

Bonnet, C. 1779–1783. *Œuvres d'histoire naturelle et de philosophie.* 8 vols. Neuchâtel: Fauché.

Cabanis, P. J. G. 1795. Note adressée aux auteurs du Magasin encyclopédique, sur l'opinion de Messieurs Oelsner et Soemmering, et du citoyen Sue, touchant le supplice de la guillotine. *Magasin encyclopédique ou Journal des Sciences, des Lettres et des Arts, 5*: 155–174.

Cabanis, P. J. G. 1823–1825. *Œuvres complètes de Cabanis.* 5 vols. Paris: Bossagne Frères & Firmin Didot.

Cabanis, P. J. G. 1981. *On the Relations between the Physical and Moral Aspects of Man, edited by George Moral.* 2 vols. Baltimore, MD, New York: Johns Hopkins University Press.

Canguilhem, G. 1998. *La connaissance de la vie* (1st ed. 1952). Paris: J. Vrin.

Chazaud, J. 1998. Cabanis before the guillotine. *Histoire des sciences médicales, 32*(1): 69–73.

Cheung, T. 2000. *Die Organisation des Lebendigen. Die Entstehung des biologischen Organismusbegriffs bei Cuvier, Leibniz und Kant.* Frankfurt am Main: Campus Verlag.

Cheung, T. 2004. Charles Bonnets allgemeine Systemtheorie organismischer Ordnung. *History and Philosophy of the Life Sciences, 26*(2): 177–207.

Cheung, T. 2006a. The hidden order of preformation: Organized bodies in the writings of Louis Bourguet, Charles Bonnet and Georges Cuvier. *Early Science and Medicine, 11*(1): 11–49.

Cheung, T. 2006b. From the organism of a body to the body of an organism: Occurrence and meaning of the word "organism" from the seventeenth to nineteenth centuries. *British Journal of the History of Science*, *39*(3): 319–339.

Cheung, T. 2008. *Res vivens. Agentenmodelle organischer Ordnung 1600–1800*. Freiburg im Breisgau: Rombach Verlag.

Cheung, T. 2010a. What is an "organism"? On the occurrence of a new term and its conceptual transformations 1680–1850. *History and Philosophy of the Life Sciences*, *32*(2–3): 155–194.

Cheung, T. 2010b. Embodied stimuli: Bonnet's statue of a sensitive agent. In O. Gal and C. Wolfe (eds.), *The Body as Object and Instrument of Knowledge* (pp. 309–331). Heidelberg, London, New York: Springer.

Cheung, T. 2014. *Organismen. Agenten zwischen Innen- und Außenwelten 1780–1860*. Bielefeld: Transcript Verlag.

Cipollini, E. M. 1998. *Analisi dei Rapports cabanisiani. Antropologia filsosofica*. Padova: Libraria Padovana Editrice.

Coleman, W. 1964. *Georges Cuvier. Zoologist. A Study in the History of Evolution Theory*. Cambridge, MA: Harvard University Press.

Corsi, P. 1988. *The Age of Lamarck*. Revised and updated. Translated by Jonathan Mandelbaum. Berkeley, CA: University of California Press.

Cuvier, G. 1798. *Tableau Élémentaire de l'histoire naturelle des animaux*. Paris: Baudouin.

Cuvier, G. 1800–1805. *Leçons d'anatomie comparée*. 5 vols. Paris: Baudouin.

Cuvier, G. 1812. *Recherches sur les ossements fossiles de quadrupèdes, où l'on rétablit les caractères de plusieurs espèces d'animaux que les révolutions du globe paroissent avoir détruites*. Paris: Déterville.

Cuvier, G. 1817. *Le règne animal distribué d'après son organisation pour servir de base à l'histoire naturelle des animaux et d'introduction à l'anatomie comparée*. 4 vols. Paris: Déterville.

Duchesneau, F. 1987. *Genèse de la théorie cellulaire*. Paris: J. Vrin.

Düsing, K. 1986. *Die Teleologie in Kants Weltbegriff*. (1st ed. 1968). 2nd ed. Bonn: Bouvier.

Duvernoy, G. L. 1833. *Notice historique sur les ouvrages et la vie de M. le B.on Cuvier*. Paris: Levrault.

Figlio, K. M. 1975. Theories of perception and the physiology of mind in the late eighteenth century. *History of Science*, *13*: 177–212.

Figlio, K. M. 1976. The metaphor of organization. An historiographical perspective on the bio-medical sciences of the early nineteenth century. *History of Science*, *14*(1): 17–53.

Gissis, S. 2010. Lamarck on feelings: From worms to humans. In O. Gal and C. Wolfe (eds.), *The Body as Object and Instrument of Knowledge* (pp. 211–239). Heidelberg, London, New York: Springer.

Greene, J. C. 1969. Biology and social theory in the nineteenth century: Auguste Comte and Herbert Spencer. In M. Clagett (ed.), *Critical Problems in the History of Science* (pp. 419–446). Madison and Milwaukee, WI, London: University of Wisconsin Press.

Guichet, J.-L. 2010. La question de l'animalité, pivot du matérialisme et de la définition de l'humain chez Cabanis. *Dix-huitième siècle*, *42*: 367–384.

Guillo, D. 2003. *Les figures de l'organisation. Sciences de la vie et sciences sociales au XIXe siècle*. Paris: P. U. F.

Haines, B. A. 1978. The inter-relations between social, biological, and medical thought, 1750–1850: Saint-Simon and Comte. *The British Journal for the History of Science, 11*: 19–35.

Hanulak, R. 2009. *Maschine – Organismus – Gesellschaft. Physiologische Aspekte eines Lebensbegriffs um 1800*. Frankfurt am Main: Peter Lang.

Henkelmann, T. 1981. *Zur Geschichte des pathophysiologischen Denkens. John Brown (1735–1788) und sein System der Medizin*. Berlin: Springer.

Holbach, P. T. de 1770. *Système de la nature*. 2 vols. London, Amsterdam: Marc-Michel Rey.

Hufeland, C. W. 1795. *Ideen über Pathogenie und Einfluss der Lebenskraft auf Entstehung und Form der Krankheiten als Einleitung zu pathologischen Vorlesungen*. Jena: Academische Buchhandlung.

Hufeland, C. W. 1799. *Bemerkungen über die Brownische Praxis*. Tübingen: J. G. Cotta.

Hufeland, C. W. 1800–1805. *System der practischen Heilkunde. Ein Handbuch für academische Vorlesungen und für den practischen Gebrauch*. 2 vols. Jena-Leipzig: Friedrich Frommann.

Hufeland, C. W. 1811. Rechenschaft an das Publikum über mein Verhältniss zum Brownianismus. *Journal der practischen Heilkunde, 32*: 3–29.

Hufeland, C. W. 1834. *Neue Auswahl kleiner medizinischer Schriften*. Vol. 1. Berlin: Veit & Comp.

Hufeland, C. W. 1851. *Enchiridion medicum oder Anleitung zur medizinischen Praxis. Vermächtniss einer funfzigjährigen Erfahrung*. (1st ed. 1836). 9th ed. Berlin: Jonas Verlagsbuchhandlung.

Hufeland, C. W. 1860. *Makrobiotik oder die Kunst das menschliche Leben zu verlängern*. (1st ed. 1796). 8th ed. Berlin: Georg Reimer.

Köchy, K. 1995. Organische Ganzheit. Die maßgeblichen Prinzipien des romantischen Organismuskonzeptes. *Biologisches Zentralblatt, 114*(2): 207–215.

Kucharczik, K. 1998. *Der Organismusbegriff in der Sprachwissenschaft des 19. Jahrhunderts*. Dissertation. Technische Universität Berlin.

Lenoir, T. 1980. Kant, Blumenbach, and vital materialism in German biology. *Isis, 71*(1): 77–108.

Lenoir, T. 1982. *The Strategy of Life. Teleology and Mechanics in Nineteenth-Century German Biology*. Chicago, IL: Chicago University Press.

Lepenies, W. 1976. *Das Ende der Naturgeschichte. Wandel kultureller Selbstverständlichkeiten in den Wissenschaften des 18. und 19. Jahrhunderts*. Munich, Vienna: Hanser.

Lesch, J. E. 1984. *Science and Medicine in France. The Emergence of Experimental Physiology, 1790–1855*. Cambridge, MA, London: Harvard University Press.

Manuel, F. E. 1956. *The New World of Henri Saint-Simon*. Cambridge, MA: Harvard University Press.

Mayer, P. M. 1993. *Christoph Wilhelm Hufeland und der Brownianismus*. Medizinische Dissertation. Johannes Gutenberg-Universität. Mainz.

Möller, H.-J. 1975. *Die Begriffe „Reizbarkeit" und „Reiz." Konstanz und Wandel ihres Bedeutungsgehaltes sowie die Problematik ihrer exakten Definition*. Stuttgart: G. Fischer.

Moravia, S. 1966. Aspetti della "Science de l'homme" nella filosofia degli "Idéologues." *Rivista critica della storia della filosofia, 21*: 398–425.

Moravia, S. 1974. *Il pensiero degli idéologues. Scienza e filosofia in Francia (1780–1815)*. Florence: La Nuova Italia.

Outram, D. 1984. *Georges Cuvier. Vocation, Science and Authority in Post-Revolutionary France*. Manchester: Manchester University Press.

Outram, D. 1986. Uncertain legislator: Cuvier's Laws of Nature. *Journal of the History of Biology*, *19*(3): 323–368.

Pfeifer, K. 1968. *Christoph Wilhelm Hufeland. Mensch und Werk. Versuch einer populärwissenschaftlichen Darstellung mit 35 Abbildungen und einem dokumentarischen Anhang*. Halle: Niemeyer.

Pfeifer, K. 2000. *Medizin der Goethezeit. Christoph Wilhelm Hufeland und die Heilkunst des 18. Jahrhunderts*. Köln: Böhlau Verlag.

Reill, P. H. 2005. *Vitalizing Nature in the Enlightenment*. Berkeley, CA: University of California Press.

Risse, G. B. 1971. *The History of John Brown's Medical System in Germany during the Years 1790–1806*. Dissertation. University of Chicago.

Robin, C. 1880. Recherches historiques sur l'origine et le sens des termes organisme et organisation. *Journal de l'anatomie et de physiologie normales et pathologiques de l'homme et des animaux*, *16*: 1–55.

Rudwick, M. J. S. 1997. *Georges Cuvier, Fossil Bones, and Geological Catastrophes: New Translations and Interpretations of the Primary Texts*. Chicago, IL: University of Chicago Press.

Schipperges, H. 1987. *Die Kunst zu leben. Ein Kommentar zu Hufelands „Makrobiotik."* Karlsruhe: Harsch.

Schmitt, S. 2009. From physiology to classification: Comparative anatomy and Vicq d'Azyr's plan of reform for life sciences and medicine (1774–1794). *Science in Context*, *22*(2): 145–193.

Schönfeld, N. 1988. *Beiträge zum ideengeschichtlichen Hintergrund der „Makrobiotik" von Christoph Wilhelm Hufeland*. Dissertation. Freie Universitaet Berlin.

Starobinski, J. 1999. *Action et réaction. Vie et aventure d'un couple*. Paris: Seuil.

Staum, M. S. 1974. Cabanis and the science of man. *Journal of the History of the Behavioral Sciences*, *10*(2): 135–143.

Staum, M. S. 1980. *Cabanis: Enlightenment and Medical Philosophy in the French Revolution*. Princeton, NJ: Princeton University Press.

Steinke, H. 2005. *Irritating Experiments. Haller's Concept and the European Controversy on Irritability and Sensibility, 1750–90*. Amsterdam, Baltimore, MD: Editions Rodolpi.

Szyfman, L. 1982. *Jean-Baptiste Lamarck et son époque*. Paris, New York: Masson.

Taquet, P. 2006. *Georges Cuvier. Naissance d'un génie*. Paris: Odile Jacob.

Trepl, L. 1994. *Geschichte der Ökologie vom 17. Jahrhundert bis zur Gegenwart*. (1st ed. 1987). 2nd ed. Frankfurt am Main: Athenäum.

Tsouyopoulos, N. 1988. The influence of John Brown's ideas in Germany. *Medical History, Supplement*, *8*: 63–74.

Vicq d'Azyr, F. 1805. *Œuvres recueillies et publiées avec des notes et un discours sur sa vie et ses ouvrages par Jacq. L. Moreau (de la Sarthe)*. 6 vols. Paris: L. Duprat-Duverger.

Warnke, C. 1998. Schellings Idee und Theorie des Organismus und der Paradigmawechsel der Biologie um die Wende zum 19. Jahrhundert. *Jahrbuch für Geschichte und Theorie der Biologie*, *5*: 187–234.

Weingarten, M. 1997. *Organismen: Objekte oder Subjekte der Evolution? Philosophische Studien zum Paradigmawechsel in der Evolutionsbiologie*. Darmstadt: Wissenschaftliche Buchgesellschaft.

Wiesing, U. 1995. *Kunst oder Wissenschaft? Konzeptionen der Medizin in der deutschen Romantik*. Stuttgart-Bad Cannstatt: Fromann-Holzboog.

Wolfe, C. T. 1999. Machine et organisme chez Diderot. *Recherches sur Diderot et sur l'Encyclopédie*, *26*: 213–231.

Wolfe, C. T. 2010. Do organisms have an ontological status? *History and Philosophy of the Life Sciences*, *32*(2–3): 195–231.

Wolfe, C. T. 2014. On the role of Newtonian analogies in eighteenth-century life science: Vitalism and provisionally inexplicable explicative devices (pp. 223–261). In Z. Biener and E. Schliesser (eds.), *Newton and Empiricism*. Oxford: Oxford University Press.

Wolfe, C. T. and Gal, O. (eds.). 2010. *The Body as Object and Instrument of Knowledge: Embodied Empiricism in Early Modern Science*. Dordrecht: Springer.

Yolton, J. W. 1991. *Locke and French Materialism*. Oxford: Clarendon.

Zammito, J. H. 1992. *The Genesis of Kant's Critique of Judgement*. Chicago, IL: University of Chicago Press.

Part III

Systems

7 Philosophy of ecology long before ecology

Kant's idea of an organized system of organized beings

Georg Toepfer

Two traditions of defining ecology

Many people think that before Darwin there was no such thing as ecology. This is true at least as far as the term itself is concerned: "Ecology" was coined by the German morphologist and evolutionary biologist Ernst Haeckel in 1866. But Haeckel only provided the word; he never worked as an ecologist. He collected animals from various places in the world but he did not investigate the interrelationships between them, as he was more interested in their morphology and phylogeny. Thus, he has a more theoretical understanding of what ecology is – or should be. In the table of biological subdisciplines in his 1866 book, ecology is considered part of *zoodynamics* or *physiology*, in contrast to *zoostatics* or *morphology* and *zoochemistry*. Under zoodynamics, it is assigned to *relations-physiology* in contrast to *conservation-physiology*, and under relations-physiology it is categorized as dealing with not *internal* relations of the parts of one body but *external* relations: the *interrelations between organisms of different species* – which is how Haeckel defines ecology in 1866.

> Ecology … the science of the interrelations [*Wechselbeziehungen*] of organisms to each other.
>
> (Haeckel 1866, II, 236)

Later, Haeckel also refers to other forms of interactions as well, and he explicitly includes *competition* and *struggle for existence* in his definition. In 1868, he gives the following definition:

> *The œcology of organisms*, the knowledge of the sum of the *relations of organisms to the surrounding outer world*, to organic and inorganic conditions of existence; the so-called "*economy of nature*," the correlations between all organisms living together in one and the same locality, their adaptation to their surroundings, their modification in the struggle for existence.
>
> (Haeckel 1868, trans. by E. Ray Lankester 1880, vol. 2, 354)

This definition of ecology as *the science of the struggle for existence* is now common. Gregory Cooper has written a book about ecology with this phrase as its title (Cooper 2003). The understanding of ecology as the science of the struggle for existence was a productive approach because it allows for ecology to become a quantitative science with rigorous research guidelines: a science of measuring the abundance and distribution of animals and plants that takes into account changes over time due to biotic and abiotic interactions. Since the mid-twentieth century, ecology has frequently been defined this way. According to Herbert Andrewartha in 1961 and Charles Krebs in 1972, ecology proper is the scientific study of (the interactions that determine) the distribution and abundance of organisms (Andrewartha 1961, 10; Krebs 1972, 4).

However, there is another tradition that instead of emphasizing *struggle* and *competition* emphasizes *interaction* and *interdependence*. Rather than focus on numbers and population size, it is primarily concerned with the *constitution* and *stability* of systems beyond the level of the individual. The roots of this tradition can be found in the view of landscapes and local assemblages of plants as unified systems, "associations" or "formations," as they were called by Alexander von Humboldt, August Grisebach and others in the first half of the nineteenth century (McIntosh 1985). The supposed unity of these ecological systems was based on different criteria, e.g., topographical boundaries, composition of the community or biotic interactions (Jax 2006). Several authors compared ecological communities to organisms, some of them being in the tradition of idealistic philosophy of nature such as Karl Friedrich Burdach, who wrote in 1842: "the various organic beings stand in interaction, are in need of each other and behave like organs of one and the same living whole" (Burdach 1842, vol. 1, 48). Three decades later, Karl Möbius provided a new word for this "community of living beings" that "depend on each other," which was "biocoenosis" (Möbius 1877, 76). But it was not before the first decades of the twentieth century that investigations of these "communities" reached the stage of solid research programs. In plant ecology this development started with the analysis of deterministic "successions" in the composition of plant communities on their way to establish a stable "climax formation," which was compared to a mature organism (Clements 1916). A similar development took place in limnology, with lakes being analyzed as "organism[s] of a higher order" (Thienemann 1918, 300) or "(superindividual) whole[s], consisting of organisms behaving with and for each other" (Thienemann 1939, 275). Until the 1950s this understanding of ecological systems was in fact the dominant view; it is manifest in the various definitions of ecology that have been given. Ecology has been labeled the doctrine of reciprocal biological relations (Drude 1906, 186), the science of communities (Shelford 1929, 2), the science of superindividual structures (systems) (Friederichs 1957, 124), and the study of the structure and function of ecosystems (Odum 1962, 108). Although he scarcely had any direct influence on this tradition of ecological systems thinking, in retrospect Immanuel Kant seemed to have provided concise ideas for this field of research long before empirical research got started. He did so with reference to Carl Linnaeus and to the even longer tradition of natural theology.

Ecological ideas in the physico-theological tradition

With his concept of the "economy of nature" (*oeconomia naturae*), Linnaeus proposes an ecological connection or *nexus* between organisms of different species. In his view this nexus is due to divine providence, even though it consists of very earthly processes, namely trophic relationships: one organism feeding on others.

Linnaeus had already described the reciprocity in the trophic relationship between organisms elsewhere before publishing his famous thesis *Oeconomia Naturae* in 1749. In an account of his travels through West-Gothland from June to August 1746, Linnaeus describes the interaction between organisms that was later taken as evidence for nature's economy. Linnaeus writes in 1747:

> When animals die they are converted into mould, the mould into plants. The plants are eaten by animals, thus forming the animals' limbs, so that the earth, transmuted into seed, then enters man's body as seed and is changed there by man's nature into flesh, bones, nerves, etc.; and when after death the body decomposes, the natural forces decay and man again becomes that earth from which he was taken.
>
> (Linnaeus, 1747, trans. by Blunt 1971, 163)

In *Oeconomia Naturae*, Linnaeus situates these observations at a more general level by stating:

> The divine wisdom has thought fit that all living creatures should constantly be employed in producing individuals; that all natural things should contribute and lend a helping hand to preserve every species; and lastly, that the death and destruction of one thing should always be subservient to the restitution of another.
>
> (Linnaeus 1749, trans. by Stillingfleet 1775, 40)

According to this view, even the adverse interactions between the organisms are manifestations of a divine plan: struggle for existence, predation, and parasitism are all interpreted as contributing to the preservation of the living community.

In his *On the Police of Nature* (1760), Linnaeus expresses this interactionist view by supplementing the "received opinion" concerning nature's teleological order – plants being there for the sake of animals – with a reverse teleology: "Attending to the order of nature, we discover that animals were created upon account of plants" (*animalia propter plantas condita esse*; Linnaeus 1760, 6, trans. by Brand 1781, 137). In support of this account Linnaeus lists some "duties of animals" toward plants, for instance with regard to nutrition and dispersal.

Thirty years later, Kant takes up these ideas in his *Critique of Judgment* (1790), which is famous for establishing systems thinking on the level of the individual organism (Quarfood 2006; Toepfer 2011). Still, at least in passing, he does extend his teleological reflection from the singular organism to a super-individual system and envisions an ecological system based on the reciprocal relationship between living beings of different species – and Linnaeus's reverse teleology may even have been important for the genesis of Kant's idea of connecting (intra-individual) organic teleology to causal reciprocity. In Paragraph 82 of his third *Critique*, Kant explicitly refers to Linnaeus and his reverse teleology – which also affects, probably most disturbingly for Kant, the position of man:

> One could also, with the Chevalier Linné, take the apparently opposite path and say that the plant-eating animals exist in order to moderate the excessive growth of the plant kingdom, by which many of its species would be choked; the carnivores exist in order to set bounds to the voraciousness of the plant-eaters; finally, humankind exists in order to establish a certain balance among the productive and destructive powers of nature by hunting and reducing the number of the latter. And thus the human being, however much he might be valued as an end in a certain relation, would in another relation in turn have only the rank of a means.
>
> (Kant 1790/1793, AA V, 427, trans. Guyer and Matthews 2000, 295)

Kant's systems thinking

Kant`s contributions to ecology have been mostly neglected so far, especially in the English-speaking world. Introductory books on the history of ecology fail to mention Kant: His name appears nowhere in Donald Worster's *Nature's Economy*, Robert McIntosh's *Background of Ecology*, or Frank Egerton's *Roots of Ecology* (Worster 1977/1994; McIntosh 1985; Egerton 2012). The same is true for other introductory books on the history of ecology.

However, there are some references to Kant's philosophy of ecology in Kant scholarship. But they are rare and most are from the first half of the twentieth century. Many are not concerned with the fundamental aspects of ecology but rather with more specific topics such as symbiosis (Stöhr 1909, 332) or autecological relationships (Roretz 1922, 74; 140).

The best known treatment of the meaning of Kant's teleology for the logic of ecology dates from 1935 and was written by the Dutch zoologist Cornelis Jakob van der Klaauw. It unfortunately appeared a few years before Kant's *Opus Postumum* was published, in which Kant expounds on the important points of his philosophy of systems ecology. Without knowledge of these texts, van der Klaauw limits his discussion to autecology, i.e., the environmental relationships of singular organisms or species.

Van der Klaauw's analysis begins with Paragraphs 63 through 84 of the *Critique of Judgment*, in which Kant stresses the foundational role of means-to-ends relations in thinking about an organism's unity. On the one hand, van der Klaauw adheres to Kant's claim that "natural purpose" (*Naturzweck*) must be restricted to individual organisms and that "nature as a whole" (*die Natur im*

Ganzen) is not presented to us "as organised" (*als organisirt*), whereas individual organisms are (Kant 1790/1793, AA V, 399, trans. Guyer and Matthews 2000, 269). Van der Klaauw thus supports the interpretation of other Kant scholars such as Emil Ungerer (1922, 125), who saw no reason, on the basis of Kant's text, to expand the system of purposes to include communities of organisms, let alone the whole of nature. On the other hand, van der Klaauw's "holism" (Trienes 1992) fostered an understanding of ecology as the science of superindividual systems in which organisms are reciprocally related to each other as means and ends, comparable to the organs within an organism: "A whole organism can (as prey) be a means to an end (a predator); it then is, at the same time, end (as a whole) and means" (van der Klaauw 1935, 565, my translation). According to van der Klaauw, these relationships result in communities of organisms with "reticulate connections" (*netzartigen Verbindungen*). Even though van der Klaauw claims that there are means-to-ends relations in such a system, he denies, in keeping with his interpretation of Kant, the whole system the status of a "natural purpose." The causal interconnection of organisms within an ecological system is considered to be the unity of a whole with teleological relations between its parts, but that unity is not a purpose in itself.

Van der Klaauw departs from Kant's own thinking insofar as he wishes to establish a hierarchical system for biology with reciprocal teleological relations at various levels. Kant, according to van der Klaauw, only had "unilinear connections" between organisms in mind, with man being nature's ultimate purpose. In contrast, van der Klaauw proposes that "in modern ecology we are allowed to read rows [of causal dependence between organisms] in both directions" (van der Klaauw 1935, 570). Organisms could be judged not only as purposes in themselves but also as functional "components" of ecological systems (van der Klaauw 1936, 235). The view of Kant's ecology being oriented in only one direction, namely toward man, was based solely on the *Critique of Judgment*.

Subsequent authors such as Heinz Heimsoeth (1940–1941) or Klaus Düsing (1968, 158–159) were able to take the *Opus Postumum* into consideration. Heimsoeth identifies in Kant's posthumous manuscripts the idea of a "whole and arrangement (*Ganzheit und Gefügtheit*) of organic bodies into 'corporations'" on the basis of a "purposiveness between different and as such independent organisms" (Heimsoeth 1940–1941, 105, 102). But these later authors only touch upon Kant's ecology in passing. This oversight stands in sharp contrast to other facets of Kant's philosophy of biology that have been studied extensively in recent years (such as his understanding of "natural history" or his theory of the organism: Goy and Watkins eds., 2014).

In the context of systems thinking, Kant's turn from *natural description* to *natural history* in the 1770s is particularly significant. Inspired by Buffon's species concept, Kant envisions a natural system of affinities between all living beings based on their genealogical relations. This is what he calls *natural history*, in contrast to *natural description*. In a famous passage from his essay on human races from 1775, extensively discussed in the last ten years by Phillip Sloan (2006) and John Zammito (2012), Kant writes:

> *Natural history*, which we are presently almost entirely lacking, would teach us about the changes in the earth's condition [*Erdgestalt*], including the changes that the creations of the earth (plants and animals) have sustained as a result of natural migrations, and about the deviations from prototype of the lineal stem species [*Stammgattung*] that have originated <in consequence of these changes>. <Natural history> would presumably lead us back from the great number of seemingly different kinds to races of just the same species and transform the very detailed scholastic system [*Schulsystem*] presently <in use> for the description of nature into a physical system [*physisches System*] for the understanding.
>
> (Kant 1775, trans. by Mikkelsen 2013, 332–333)

In the 1770s, Kant's program for natural history aims at the transformation of a merely *scholastic system* of classification into a *natural system* of real relations. He pursues this by proposing temporal dimensions meant to establish a physical system of living beings on the basis of their common descent. This is *natural history* in its proper, *temporal* sense. As Kant explicitly states, a *natural* division is "based upon the common stem, which divides animals according to kinship from the standpoint of generation" – in contrast to a scholastic division that divides things up according to *similarities* (Kant 1775, trans. by Sloan 2006, 634). Kant's evolutionary and, at the time, utopian vision was not well received. Georg Forster, professor of natural history at Kassel, claimed that such a "history" of nature is "a science for gods and not for men" (Forster 1786, 80). Nevertheless, in his debate with Kant, Forster was the very first to use the term "phylogenetic tree" (*Stammbaum*) to refer to species descent. Kant, however, became more cautious with old age and in the *Critique of Judgment* calls his own phylogenetic hypothesis a "daring adventure of reason" (§80; Zöller 2011; Zammito 2012).

Kant would maintain his vision for a natural system based on physical relations (actually *causal relations*) not only in *temporal* terms but also in *spatial* terms – and this led him to his philosophy of ecology.

In his *Lectures on Physical Geography*, which he delivered 49 times during his university career from 1756 to 1796, Kant insists foremost on the point that a true system of nature has yet to be established:

> [T]he systems of nature that have been drawn up so far are probably more properly called aggregations of nature, for a system presupposes an idea of the *whole*, from which the diversity of things is derived. Actually, we do not yet have *systema naturae*. In the so-called systems of this type available at present, the things are simply put together and arranged in series.
>
> (Kant 1802, trans. by Reinhardt 2012, 448; cf. Kant 1756–1759, 10)

A few sentences later, Kant makes clear that there are two ways to set up a system on the basis of physical relations, either through history or geography:

> History and geography extend our knowledge in relation to time and space. History concerns the events that have taken place one after another in time. Geography concerns phenomena that occur simultaneously in space.
> (Kant 1802, trans. by Reinhardt 2012, 449; cf. Kant *c.*1775, 160)

Kant's idea is that in order to establish physical systems one has to presuppose the idea of the *whole*. Wholes can be realized in time or space. They are thus established by the frameworks of *history*, or to be more precise *historiography*, and *geography*.

However, Kant does not grant historiography and geography equal status as disciplines. He sees geography as primarily descriptive, while the aim of historiography is to provide explanations or at least a coherent account of events. In the version of Kant's lecture on "Physical Geography" that was revised by Friedrich Theodor Rink, it is simply stated, "History is a narrative [*Erzählung*], but geography is a description" (Kant 1802, trans. by Reinhardt 2012, 449–450). Even as a descriptive science, geography should be guided, according to Kant, by the idea of wholes or systems and their spatial components (Fisher 2007, 104).

Kant's perspectives on historiography (for nature) and geography as two approaches that are based on the idea of systems in time and space respectively can be illustrated with graphical models that were developed much later but which capture Kant's ideas very nicely. One is the phylogenetic tree, which shows successive speciation events in time and allows for a nested classification of taxonomic groups. In these phylogenetic diagrams, we find wholes or natural systems that rely on causal chains going back in time. Here groups of organisms or species are delimited by sequences of bifurcating events, resulting in a nested hierarchy of classification. In spatial terms, the corresponding graphical representation for geographical systems would be a diagram of an ecological system in which the closure of the system is represented by the cyclicity of the causal (trophic) interactions between organisms of different species. The ecological diagram presents a whole or natural system based on simultaneous interactions. It corresponds to Kant's idea of geographical systems – another "daring adventure of reason" because these systems cannot be experienced directly. This delimitation of physical systems in one area leads the way to Kant's ecology, or, to be more precise, to his philosophical *idea* of a theoretical basis *for* ecology.

Kant's ecology: "organizing systems of organized bodies"

A few years before his death, in the summer of 1799, Kant expands on the idea of how to establish *physical systems* on the basis of causal interactions between organisms in one area. In fact, he formulates the principle of an *ecosystem*, in the sense of a system of mutually dependent organisms from different species – in Kant's words, "the natural system of the purposive relation between different species which exist for each other's sake" ("das Natursystem in dem zweckmäßigen Verhältnis verschiedener Arten deren eine um der anderen Willen da ist"; Kant, OP AA, 1799, XXI, 566). This natural system is based on "the organizing of systems of organized bodies" ("Die Organisirung der Systeme von organisirten Körpern"; ibid.). In other words, Kant is suggesting a supersystem, a second-order organization, the organization of organisms into systems at a higher level.

During the summer of 1799, Kant entertains many versions of how to best formulate these superorganismic organizations. As can be seen in Kant's manuscripts which are full of crossing outs and corrections, this seems to be a playful, if perhaps painful activity because he often rephrases what he wrote.

Kant speaks of an "organization of a whole consisting of different species of organized beings serving each other and their preservation" (OP AA, 1799, XXII, 300) and of an "organization of a system of organized beings, e.g., deer for the wolf, mosses for the tree, soil for the crop" (OP AA, 1799, XXII, 505). He claims that "Nature does not only organize matter into bodies but these in turn into corporations [*Corporationen*] which for their part have relationships of reciprocal purposiveness (one being there for the sake of the other)" (OP AA, 1799, XXII, 506). Finally, he envisions "systems of living bodies in as much as one belongs as a part to the life of the other (e.g., reindeer and moss or sheep and wolf)" (OP AA, 1799, XXII, 534). These are clearly ecological relationships. The mention of reindeer here makes it likely that he received some of his insights from Linnaeus. But, as is always the case in Kant, he tries to hone the ideas and turn empirical observations into fundamental principles, which, according to him, were the building blocks of science.

It is striking, however, that Kant does not mention reciprocal relationships between organisms in his concrete examples. Instead, he only lists unidirectional cases ("deer for the wolf, mosses for the tree, soil for the crop"; see also Düsing 1968, 158–159). Nevertheless, Kant's theoretical phrasing makes it clear that he had "reciprocal purposiveness (one being there for the sake of the other)" in mind. The interactions between plants and animals preserve "a certain balance among the productive and destructive powers of nature" (Kant 1790/1793, AA V, 427, trans. Guyer and Matthews 2000, 295).

Finally, Kant extends this systems-theoretical perspective to the whole of nature:

> [The] organizing force [of our living, fertile globe] has so arranged for one another the totality of the species of plants and animals, that they, together, as members of a chain, form a circle (man not excepted). That they require each other for their existence, not merely in respect of their nominal character (similarity), but their real character (causality) – which points in the direction of a world organization (to unknown ends) of the galaxy itself.
>
> (Kant, OP AA, XXI, 570, trans. Förster 1993, 86)

The idea of all species of organisms being connected to one integrated organization points toward the concept of an *ecosystem*, or of *gaia*, which understands the entire earth as involved in a circle of causal dependencies. Inorganic parts of the earth are also included in this organization. In his *Opus Postumum*, Kant calls the earth an "organized world body [*organisirten Weltkörper*] even with respect to its inorganic parts or also organic for mutual consumption determined organic bodies" (Kant, OP AA, XXII, 504; cf. 276; XXI, 196; 215). Therefore, for Kant, "organized bodies" exist on various levels as individual organisms, particular ecological systems, and the entire earth.

This extended view contradicts Kant's previous claim in his *Critique of Judgment*, where he states that "there is only one external purposiveness which is connected with the internal purposiveness of organization, and yet serves in the external relation of a means to a purpose" which is "the organization of the two sexes in relation to one another for the propagation of their kind" (Kant 1790/1793, AA V, 425, trans. Guyer and Matthews 2000, 293) Kant admits at this juncture that it would be "conformable to Reason" to make "an objective purposiveness in the variety of the genera of creatures and their external relations to one another" a principle of cognition in order "to conceive in these relations a certain organization and a system of all natural kingdoms according to final causes" (ibid., 349). But, Kant continues, "experience seems flatly to contradict the maxims of Reason" (ibid.). Besides the empirical reasons against the existence of superindividual organizations in nature, Kant also provides a theoretical argument: The possibility of such a "system" of organisms would presuppose an "ultimate purpose of nature," which Kant thinks could be "put nowhere else but in man" (ibid.).

In 1799 Kant is still of the opinion that man is the only ultimate purpose of nature. But, in contrast to his position in the *Critique of Judgment*, he now regularly speaks of superindividual organizations in nature. Apparently what had changed was his concept of organization: Whereas in the third *Critique* an organization is conceivable only as a natural purpose, in the *Opus Postumum* it seems to be possible as a result of the mere interaction of organized subsystems (the organisms) if they are linked through "reciprocal purposiveness (one being there for the sake of the other)." At this stage every system that consists of mutually dependent parts seems to be an organization for Kant, although it need not be a natural purpose in itself. Consequently, even in the *Opus Postumum*, Kant does not claim that the parts of superindividual organizations exist for the sake of the whole – as he does in the *Critique of Judgment* with respect to the parts of an individual organism (Tanaka 2004, 291).

Another difference to Kant's thinking in the *Critique of Judgment* is that in the *Opus Postumum* the organic systems are determined in explicitly causal terms. The level of the system is not relativized as only being given in reflective judgment – the term "reflective judgment," which plays a central role in the third *Critique*, does not appear once in the *Opus Postumum* (Tanaka 2004, 277). Some scholars conclude from this that the concept of purpose in the *Opus Postumum* had obtained a constitutive status for the knowledge of natural systems (Tanaka 2004, 280; Ingensiep 2009, 106; Basile 2013, 320). However, contrary to this view, Kant insists, even in his *Opus Postumum*, that the whole was a "mere idea" and that the reality of organic wholes could not be secured on an a priori basis (Kant OP AA, XXI, 210). Following Heimsoeth (1940–1941, 108), it can be said that Kant's thoughts in the *Opus Postumum* are not " 'dogmatic' constructions" but ideas on reflective judgment. In support of this view, it is also revealing that Kant describes final causes as a "guiding thread for research" (*Leitfaden für die Naturforschung*) in order to see if and how far they, too, form a system (OP AA, XXI, 184). As a guiding thread the idea would not interfere with causal analyses but provide a tool for *presenting* the manifold as an ordered system.

Kant's talk of a "guiding thread" in this context suggests a parallel reading for the status of the idea of superindividual (ecological) systems (and genealogical taxonomic classes) on the one hand and the ideas of historiography on the other hand. The representation of nature according to the ideas of geographical systems and historiographical classes would parallel the representation of human history according to the "guiding thread" of a "cosmopolitan intent" that involves man's "cultivation, civilization and moralization," an idea which serves us "for exhibiting an otherwise planless aggregate of human actions, at least in the large, as a *system*" (Kant 1784, AA VIII, 29, trans. A. W. Wood 2007, 118).

Kant does not elaborate on how to connect the idea of natural systems of organisms (i.e., the cyclical model of ecology) with man's position as the ultimate purpose of nature (i.e., the linear model of cultural historiography). Given that he explicitly cites Linnaeus's idea of a reverse teleology and the need for cyclical dependencies to delimit superindividual (ecological) systems, it seems implausible that the only purpose of the ecological perspective is to uphold the (historiographical) idea of the perfection of human beings (cf. van den Berg 2014, 257). Instead, the numerous passages in which Kant refers to superindividual systems in nature without including man in the picture suggest that he understands these ideas as providing an additional (ecological) perspective in their own right.

This view finds support in the fact that Kant repeatedly introduces the systems-theoretical perspective on functionally closed superindividual systems with the aim of differentiating between fundamental biological subdisciplines and the intentions of biological inquiries. Here are two of these classifications.

In the "division of the natural system" (*Eintheilung des Natursystems*), Kant differentiates between two categories each having two subcategories. The first category is "natural systems of organic bodies" within one species, and they include, first, associations by proximity, meaning social behavior, and, second, sequential reproduction of the same genus with the other sex, meaning the *genealogical series* within one species. The second category concerns a "natural system in the purposive relation of different species being there for each other's sake." This divides into, on the one side, *spatial juxtaposition*, "side by side," as Kant says (and which I would assign to *ecology*), and, on the other side, *temporal succession*, in Kant's words "the sequence of world epochs one after the other," which corresponds to the modern concept of *evolution* (Kant, OP AA, 1799, XXI, 566).

A second classification of research perspectives in which ecology appears prominently takes the enumeration of "moving" or "animal powers" (*animalische*

Potenzen) as its starting point. Here Kant differentiates between four levels in a hierarchical system:

> [Z]oonomy rests on the employment of four animal powers [*animalische Potenzen*]. (1) on *nervous power* as a principle of excitability (*incitabilitas Brownii*); (2) on *muscular power* (*irritabilitas Halleri*); (3) on a *force which preserves* all the organic forces of nature as a constant alteration of the former two; of which *one* phenomenon is heat; (4) on the *organization of a whole of organic beings of different species*, for each other, serving for the species' preservation.
>
> (Kant, OP AA, XXII, 300)

Here the topic of ecology, as "the organization of a whole of organic beings of different species," appears aligned with fundamental organic forces. Kant's view in the summer of 1799 is that the ecological account of integrating organisms into superindividual systems is one fundamental aspect of biology. Its basic principle consists of a single *idea* or *logic*: the organization of nature into wholes of interdependent parts.

Although at the end of his life Kant was interested in matters of ecology, he was unaware of important work by empirical researchers in this emerging field. Apparently he did not pay much attention to the development of ecological models for the circulation of matter, which started in the 1770s with experiments performed by Joseph Priestley and the chemical theories of Antoine Laurent de Lavoisier. These experiments showed that plants reversed the effects of animal breathing on the air and that plants and animals can survive much longer when put together in closed vessels than when they were separated from each other. One of the reasons for Kant's neglect of these findings might be that he had a strong inclination toward phlogiston theory (Stark 2013, 246).

Kant's contribution to ecology

Some recent Kant scholars are of the opinion that Kant's contributions to the constitution of biology were in the past rather overrated. Werner Stark, for example, claims that Kant "missed" the beginning of biology as a science *sui generis* (Stark 2013, 243; see also Zammito 2006). In contrast to assessments of this kind, it might be better to stress that Kant surely took note of the results of empirical investigations and theoretical discussions in the emerging field of biology. He reflected on them and developed concepts for them. At the beginning of the nineteenth century, his contributions were taken up by empirical researchers and were conducive to the constitution of biology. Kant's work here focused on fundamental concepts for the new field, such as the definition of organism as a system of interdependent and functionally related components, the concept of self-organization, the biological concept of species, which Kant clearly distinguishes from the morphological concept and for which he provides a precise definition following Buffon (Kant 1775, 45–46), and the plan for a "natural system" as a classification based upon genealogical relationships (see above). The dense phrasing that Kant develops for (ecological) systems of interacting organisms in one area could also be added to this line of conceptual work that was potentially fruitful for biology.

But, in striking contrast to the extensive epistemological discussions of "organized beings" in the *Critique of Judgment*, Kant does not discuss methodological questions about the epistemic status of superorganismic organizations in the *Opus Postumum*. He writes about these systems in many passages as if they could be observed in nature. This is also how early Romantic authors conceptualize grand organizations. In 1793, Kielmeyer, for instance, envisions a system of interconnected effects between organisms of different species that combine into a "huge machine of the organic world" (Kielmeyer 1793, 5); in 1798 Schelling considers nature to be a "general organism" (1798, 257); and Hegel speaks of the "geological organism of the earth" (Hegel 1830, II, 361). These were factual Romantic claims. According to Kant, the circumstances are different at least in part because, as he insists in the *Opus Postumum*, these natural organizations are *ideas*.

Ecological organizations – or, in Kant's words, "the organization of a whole of organic beings of different species, for each other, serving for the species' preservation" – are not given in experience for Kant. They are models present only in teleological reflection, analogous to organisms. They are ideas that organize our experience of spatial (geographical) phenomena in the same way that ideas of systemic unity (for nature) or cultivation (for cultural history) organize our investigations of history. These ideas have a specific epistemic role, namely to determine the direction of empirical research.

However, in the case of Kant's thoughts on the constitution of ecological systems, his idea did not guide or stimulate any empirical investigations; rather, it encouraged romantic speculations on the "general organism" of nature or the "geological organism of the earth." It thus seems that Kant's philosophy of ecology was too early to be effective for science. Ecological methods for delimiting ecological systems or for proving mutual dependencies between organisms of different species had not yet been developed. Kant's idea of an organized system of organized beings can thus be seen as a straightforward advancement of his theoretical philosophy of organized systems in nature – but it is only in retrospect that his wording can be understood as expressing the quintessence of ecological systems thinking.

References

Andrewartha, H. G. 1961. *Introduction to the Study of Animal Populations*. Chicago, IL: University of Chicago Press.

Basile, G. P. 2013. *Kants Opus Postumum und seine Rezeption*. Berlin: De Gruyter.

Berg, H. van den. 2014. *Kant on Proper Science. Biology in the Critical Philosophy and the Opus Postumum*. Springer, Dordrecht.

Burdach, K. F. 1842. *Blicke ins Leben. Comparative Psychologie*. 2 vols. Leipzig: Voß.

Clements, F. E. 1916. *Plant Succession. An Analysis of the Development of Vegetation.* Washington, DC: Carnegie Institution.

Cooper, G. 2003. *The Science of the Struggle for Existence. On the Foundations of Ecology*. Cambridge: Cambridge University Press.

Drude, O. 1906. The position of ecology in modern science. In H. J. Rogers (ed.). *[International] Congress of Arts and Science St. Louis 1904*, vol. 5 (pp. 179–190). St. Louis: Universal Exposition.

Düsing, K. 1968. *Die Teleologie in Kants Weltbegriff*. Bonn: Bouvier.

Egerton, F. N. 2012. *Roots of Ecology. Antiquity to Haeckel*. Berkeley, CA: University of California Press.

Fisher, M. 2007. Kant's explanatory natural history: Generation and classification of organisms in Kant's natural philosophy. In P. Huneman (ed.) *Understanding Purpose Kant and the Philosophy of Biology* (pp. 101–121). Rochester, NY: University of Rochester Press.

Forster, G. 1786. Noch etwas über die Menschenraßen. *Teutscher Merkur* 1786 (4th Quart.): 57–86.

Friederichs, K. 1957. Der Gegenstand der Ökologie. *Studium Generale, 10*: 112–144.

Goy, I. and Watkins, E. (eds.). 2014. *Kant's Theory of Biology*, Berlin: De Gruyter.

Haeckel, E. 1866. *Generelle Morphologie der Organismen*, 2 vols. Berlin: Reimer.

Haeckel, E. 1868. *Natürliche Schöpfungsgeschichte*, English trans.: *The History of Creation*, trans. by E. Ray Lankester. 2 vols. New York: Appleton, 1880.

Hegel, G. W. F. 1830. *Enzyklopädie der Philosophischen Wissenschaften im Grundrisse*. In Hegel, *Werke*, vols. 8–10. Frankfurt am Main 1986.

Heimsoeth, H. 1940–1941. Kants Philosophie des Organischen in den letzten Systementwürfen. *Blätter für deutsche Philosophie, 14*: 81–108.

Ingensiep, H. W. 2009. Probleme in Kants Biophilosophie. Zum Verhältnis von Transzendentalphilosophie, Teleologiemetaphysik und empirischer Bioontologie bei Kant. In E.-O. Onnasch (ed.) *Kants Philosophie der Natur. Ihre Entwicklung im Opus Postumum und Ihre Wirkung* (pp. 79–114). Berlin: De Gruyter.

Jax, K. 2006. Ecological units: Definitions and application. *Quarterly Review of Biology, 81*: 237–258.

Kant, I. 1756–1759. [*Vorlesungen zur Physischen Geografie*] [*Lectures on Physical Geography*] (Manuscript Holstein). In W. Stark (ed.) (2009) *Kant's Gesammelte Schriften*, vol. XXVI.1. Berlin: De Gruyter.

Kant, I. [*c.*1775]. [*Vorlesungen zur Physischen Geographie*] (*Vorlesungsmitschrift Manuskript Kaehler*), ed. by F. T. Rink. In Königlich Preußische Akademie der Wissenschaften (ed.) (1923) *Kant's Gesammelte Schriften*, vol. IX (pp. 151–436). Berlin: De Gruyter.

Kant, I. 1775. *Von den Verschiedenen Racen der Menschen. English trans.: On the different human races: An announcement for lectures in physical geography in the summer semester 1775*. In J. M. Mikkelsen (ed.) (2013) *Kant and the Concept of Race: Late Eighteenth-Century Writings* (pp. 41–54, 332–333). Albany, NY: State University of New York Press.

Kant, I. 1784. *Idee zu einer allgemeinen Geschichte in weltbürgerlicher Absicht*. English trans.: Idea for a universal history with a cosmopolitan aim, trans. by A. W. Wood. In G. Zöller and R. B. Louden (eds.) (2007) *Immanuel Kant: Anthropology, History, and Education* (pp. 107–120). Cambridge: Cambridge University Press.

Kant, I. 1790/1793. *Kritik der Urteilskraft*. English trans.: *Critique of the Power of Judgment*, trans. by P. Guyer and E. Matthews. Cambridge: Cambridge University Press, 2000.

Kant, I. 1802. *Vorlesung zur Physischen Geographie*. English trans.: *Physical Geography*, trans. by O. Reinhardt. In E. Watkins (ed.) (2012) *Immanuel Kant: Natural Science* (pp. 434–679). Cambridge: Cambridge University Press.

Kant, I. (OP AA). *Opus Postumum*. In A. Buchenau (ed.) (1936–1938) *Kant's Opus Postumum*, 2 vols. Berlin: De Gruyter.

Kant, I. (OP Förster). *Opus postumum*, trans. by E. Förster. In E. Förster (1993) *Immanuel Kant: Opus Postumum*. Cambridge: Cambridge University Press.

Kielmeyer, C. F. 1793. *Ueber die Verhältniße der Organischen Kräfte unter einander in der Reihe der verschiedenen Organisationen, die Gesetze und Folgen dieser Verhältniße*. Stuttgart.

Klaauw, C. J. van der. 1935. Die Bedeutung der Teleologie Kants für die Logik der Ökologie. *Sudhoffs Archiv für Geschichte der Medizin und der Naturwissenschaften*, *27*: 516–588.

Klaauw, C. J. van der 1936. Zur Aufteilung der Ökologie in Autökologie und Synökologie, im Lichte der Ideen als Grundlage der Systematik der zoologischen Disziplinen. *Acta Biotheoretica*, *2*: 195–241.

Krebs, C. J. 1972. *Ecology. The Experimental Analysis of Distribution and Abundance*. New York: Harper and Row.

Linnaeus, C. 1747. Wästgöte-Resa. English trans.: Blunt, W. (1971) *The Compleat Naturalist: A Life of Linnaeus*. London: William Collins and Sons.

Linnaeus, C. 1749. *Œconomia naturae*. English trans.: *The Œconomy of Nature*. In B. Stillingfleet (ed.) (1775) *Miscellaneous Tracts Relating to Natural History, Husbandry, and Physick* (pp. 37–129). London: Dodsley, Baker and Leigh, Payne.

Linnaeus, C. 1760. *Politia naturae*. English trans.: *On the Police of Nature*. In: *Select Dissertations from the Amoenitates Academicae*, trans. by F. J. Brand (pp. 129–166). London: Robinson and Robson, 1781.

McIntosh, R. P. 1985. *The Background of Ecology. Concept and Theory*. Cambridge: Cambridge University Press.

Möbius, K. 1877. *Die Auster und die Austernwirtschaft*. Berlin: Wiegandt, Hempel & Parey.

Odum, E. P. 1962. Relationship between structure and function in the ecosystem. *Japanese Journal of Ecology*, *12*: 108–118.

Quarfood, M. 2006. Kant on biological teleology: Towards a two-level interpretation. *Studies in History and Philosophy of Biological and Biomedical Sciences*, *37*: 735–747.

Roretz, K. 1922. *Zur Analyse von Kants Philosophie des Organischen*. Vienna: Hölder.

Schelling, F. W. J. 1798. *Von der Weltseele. Eine Hypothese der höheren Physik zur Erklärung des allgemeinen Organismus*. In K. T. Kanz and W. Schiechle (eds.) (2000) *Friedrich Wilhelm Joseph Schelling. Historisch-Kritische Ausgabe*, vol. I, 6. Stuttgart; Frommann-Holzboog.

Shelford, V. E. 1929. *Laboratory and Field Ecology*. Baltimore, MD: Williams & Wilkins.

Sloan, P. 2006. Kant on the history of nature. The ambiguous heritage of the critical philosophy for natural history. *Studies in the History and Philosophy of the Biological and Biomedical Sciences*, *37*: 627–648.

Stark, W. 2013. Naturgeschichte bei Kant. *Akten des XI. Internationalen Kant-Kongresses*, Pisa 2010, vol. 5 (pp. 233–247), Berlin: De Gruyter.

Stöhr, A. 1909. *Der Begriff des Lebens*. Heidelberg: Winter.

Tanaka, M. 2004. *Kants Kritik der Urteilskraft und das Opus Postumum. Probleme der Deduktion und Ihre Folgen.* Dissertation. University of Marburg.

Thienemann, A. 1918. Lebensgemeinschaft und Lebensraum. *Naturwissenschaftliche Wochenschrift N. F., 17*: 281–290; 297–303.

Thienemann, A. 1939. Grundzüge einer allgemeinen Ökologie. *Archiv für Hydrobiologie, 35*: 267–285.

Toepfer, G. 2011. Kant's teleology, the concept of the organism, and the context of contemporary biology. *Logical Analysis and History of Philosophy, 14*: 107–124.

Trienes, R. 1992. Holism and Kantian teleology in C. J. van de Klaauw's structuralization of oecology. *Acta Biotheoretica, 40*: 11–22.

Ungerer, E. 1922. *Die Teleologie Kants und ihre Bedeutung für die Logik der Biologie.* Berlin: Borntraeger.

Worster, D. 1977/1994. *Nature's Economy. A History of Ecological Ideas.* Cambridge: Cambridge University Press.

Zammito, J. 2006. Teleology then and now: The question of Kant's relevance for contemporary controversies over function in biology. *Studies in History and Philosophy of Biological and Biomedical Sciences, 37*: 748–770.

Zammito, J. 2012. Should Kant have abandoned the "daring adventure of reason"? – The interest of contemporary naturalism in the historicization of nature in Kant and idealist Naturphilosophie. *International Yearbook of German Idealism, 8*: 130–164.

Zöller, G. 2011. Eine "Wissenschaft für Götter." Die Lebenswissenschaften aus der Sicht Kants. In C. F. Gethmann (ed.) *Lebenswelt und Wissenschaft* (pp. 877–892). XXI. Deutscher Kongreß für Philosophie, Kolloquienbeiträge.

8 "All is leaf"

Goethe's plant philosophy and poetry

Ina Goy

When Johann Wolfgang Goethe developed his theory of plant morphology and metamorphosis, a centuries-long plant research already existed. This plant research at times had been anthropocentric, concerned with the human uses of plants, at other times botanical, concerned with plants in their own right. The earliest plant research in Europe consisted in an oral transfer of agricultural plant knowledge from generation to generation, until Aristotle and Theophrastus founded botany as a science in ancient Greece, and developed theories in the fields of plant geography, morphology, and physiology, plant nutrition, growth, and plant reproduction. In the Middle Ages most of this early scientific knowledge came to a standstill or even got lost. In the early modern period, plant studies, again, became subordinate to agricultural or medical purposes, and were collected in the form of herbals and books, which mostly consisted of compilations of classical medical texts. Around the middle of the seventeenth century herbals slowly transformed into floras, in which botanists collected knowledge of native plants of local regions and again tried to relate plants to one another and not to man. The invention and improvement of the microscope supported this revival of botany as a science in its own right and led the foundation of subjects like plant anatomy and experiments in plant physiology. With the expansion of trade and exploration beyond Europe, many new plants were discovered, and necessitated an increasingly rigorous process of naming, description, and classification of plants. And so it is safe to say that Goethe's plant morphology and metamorphosis was part of a rich and multifaceted state of botany in the late eighteenth and early nineteenth centuries.

1 The primal plant and its organ, the leaf

In *The Metamorphosis of Plants* (1790) and the related didactic poem (1798), Goethe describes the generation and development of plants as six metamorphoses of the primal plant (*Urpflanze*) and its organ, the leaf (*Blatt*). In a first step, I will try to analyze the nature of the primal plant and its organ, the leaf. Is it an idea or does it consist in matter? If it is an idea, is it a Platonic or Kantian idea? Is it the idea of all plants (*the* plant, its genus), the idea of a particular species of plants, or the idea of the individual plant? If it is matter, which kind of matter? In order to answer these questions, I will first consider Goethe's diary notes during his journey in Italy (Section 1.1). Goethe always emphasizes that observation gives him access to the primal plant and its organ, the leaf, but both notions differ in different passages of the text: while some of Goethe's earlier notes would allow a more concrete (materialist, empirical) interpretation of the primal plant and its organ, the leaf, his later depictions of both terms tend to be more abstract (conceptual, rationalist). Another ambiguity of both notions in Goethe's notes is that the primal plant and its organ, the leaf, sometimes seem to be unchangeable principles of change, sometimes changeable as principles themselves.

Then I will turn to the literature (Section 1.2). While scholars rarely consider the primal plant and its organ, the leaf, a material principle (Section 1.2.1), they more often suggest formal (Section 1.2.2), or material formal readings (Section 1.2.3). A small group of scholars describe the primal plant and its organ, the leaf, as a dynamical, mutable principle (Section 1.2.4). I will argue that the ambiguities in Goethe's own notes provoke the different readings of the nature of the primal plant and its organ, the leaf. The most plausible reading in my view is that Goethe's primal plant and its organ, the leaf, is a formal, changeable principle, which expands and contracts in plant matter (Section 1.2.5).

1.1 The primal plant and its organ, the leaf, in Goethe's notes

I will begin with an analysis of significant diary notes and texts that Goethe wrote during his journey to Italy in 1786/1787, in which he describes his discovery of the notion of the primal plant and its organ, the leaf. In the earliest of these notes, on September 27, 1786, Goethe writes about the vegetation of the botanical gardens in Padua:

> Here where I am confounded with a great variety of plants, my hypothesis that it might be possible to derive all plant forms from one primal plant becomes clear to me and more exciting. Only when we have accepted this idea it will be possible to determine the genera and species exactly. So far this has, I believe, been done in a very arbitrary way.
>
> (HA 11.60)

In these lines Goethe states that the observation of the surrounding manifoldness of plants in the gardens of Padua reminds him of the idea of the primal plant. The passage is ambiguous with regard to the nature of the primal plant; Goethe seems to either indicate that he derives the concrete manifoldness of plant forms around him from the idea (*Gedanke*) of a formerly existing, concrete ancestral plant, or that he derives it from an abstract eternal idea of the primal plant. That is, Goethe's remarks are not entirely clear with regard to whether the primal plant once existed and was the first most complex concrete member and originator of the genus of plants or whether it consists in one abstract (Platonic) idea of *the* genus of all plants. Goethe claims that the primal plant allows for the determination of all kinds of plants and insists on its singularity and uniqueness.

Another remark on what Goethe calls his "botanical obsession" (the search for the primal plant) follows on February 19, 1787, while Goethe was in Rome: "My botanical obsessions reinforce in view of all these [circumstances], and I am on the way to explore new beautiful conditions, under which nature develops the most manifold out of something so simple" (HA 11.175). Goethe reiterates that it is the observation of the surrounding nature which reinforces his botanical studies, and he emphasizes the singularity (*Einfachheit*) and uniqueness of the principle, which grounds the manifoldness of plants; it is something "out of which" the large diversity of plants develops.

Two months later, on April 17, 1787, Goethe notes about the plants in the botanical gardens in Palermo on Sicily:

> Here where, instead of being grown in pots under glass as they are with us, plants are allowed to grow freely in the open fresh air and fulfill their natural destiny, they become more intelligible. Seeing such a variety of new and renewed forms, my old fancy suddenly came back to mind: among this multitude might I not discover the primal plant? There certainly must be one. Otherwise, how could I recognize that this or that form was a plant if all were not built on the same basic model?
>
> (HA 11.266, see also 374)

In this note Goethe uses the term primal plant (*Urpflanze*) and describes it as the singular basic model (*ein Muster*), which is the ground of all forms of plants. The remark shows the same and even more ambiguity with regard to the nature of the primal plant as Goethe's notes on September 27, 1786. Goethe seems to indicate that the primal plant is either something that he tries to discover among the present plants that surround him in the botanical gardens of Palermo, or something that he tries to derive as the idea of an ancestral plant from the surrounding plants, or tries to abstract as an idea from all surrounding existing plants. Also in this note Goethe calls the primal plant a model, which seems to indicate a rather abstract principle.

Having arrived in Segesta, Goethe further notes on April 20, 1787: "Seeing fresh fennel, I realize the difference between the lower and the upper leaves, and it is always only the same organ, which develops out of simplicity into the manifoldness" (HA 11.271). In this note Goethe's focus shifts slightly from the primal plant to its organ (*Organ*), the leaf (*Blatt*). Goethe repeats that the generative principle of the manifoldness of plants is singular und unique. But in these lines it is the leaf and not the primal plant itself that is generative for concrete leaf-like parts of the plant. And it is again a concrete observation that leads Goethe to his insights.

On May 17, 1787, Goethe writes in Naples to Johann Gottfried Herder:

> I am very close to the reproduction and organization of plants, and … it is the simplest thing imaginable. This climate offers the best possible condition for making observations. To the main question – where the germ is hidden

> – I am quite certain I have found the answer; to the others I already see a general solution, and only a few points have still to be formulated more precisely. The primal plant is going to be the strangest creature in the world, which Nature herself shall envy me. With this model and the key to it, it will be possible to go on forever inventing plants and know their existence is logical; that is to say, if they do not actually exist, they could, for they are not the shadow phantoms of vain imagination, but possess an inner necessity and truth. The same law will be applicable to all other living organisms.
>
> (HA 11.323–11.324, see also 375)

In this statement on the primal plant, Goethe names it a simple model (*Modell*) and key to the understanding of the reproduction and organization of plants, and emphasizes its singularity and simplicity. Goethe also discovers now that the primal plant allows the anticipation of future plants whose forms are possible but do not exist yet, and that something similar to the primal plant is the underlying generative principle of other organisms, such as animals.

On the same day, May 17, 1787, Goethe writes in a report of his botanical studies:

> [I]n the organ of the plant, which we usually speak of as a leaf, the true Proteus is hidden, which can conceal and reveal itself in all forms. Forward and backward, the plant is always only a leaf, so inseparably united with a future germ, that one must not think one without the other. To grasp such a concept, to bear it, to find it in nature, is the task which puts us in an embarrassingly sweet state.
>
> (HA 11.375)

These are the lines in which Goethe formulates his famous claim that in a plant everything is leaf, and names the leaf a concept (*Begriff*), but also the Proteus of all forms of plants. In Greek mythology, Proteus (*Πρωτεύς*) is an early sea-god or god of rivers and oceanic bodies of water. Homer calls him the "old man of the sea," others the god of "elusive sea change," a description that points to the constantly changing nature of the sea and of liquid quality of water in general. Proteus can foretell the future, but will change his shape to avoid having to. From this feature of Proteus comes the adjective protean, with the general meaning of "versatile," "mutable," "capable of assuming many forms," with positive connotations of flexibility, versatility, and adaptability. When Goethe calls the leaf a Proteus he seems to indicate that it is a single changeable principle of plant generation and development that can take on many different forms. The interesting new feature of this association is that Goethe considers the generative principle itself changeable. It is what transforms itself into the manifoldness of concrete parts of the plant. But Goethe also calls the notion of a leaf a "concept" and emphasizes its conceptual, formal nature.

In the *History of His Botanical Studies* (*Geschichte seiner botanischen Studien* 1817b) Goethe further remarks retrospectively about his journey to Italy:

> Because they [the plants] may be grouped under one concept, it gradually became clear to me that the concept could also be valid in a higher sense: a challenge which hovered in my mind at the time in the sensible form of a supersensible primal plant. I traced the variations of all forms as I came upon them. In Sicily, the final goal of my journey, the conception of the original identity of all plant parts had become completely clear to me; and everywhere I attempted to pursue this identity and to catch sight of it again.
>
> (HA 13.164)[1]

In this remark Goethe speaks of the sensible form of a supersensible primal plant (*die sinnliche Form einer übersinnlichen Urpflanze*), which, again, is a concept (*Begriff*) for him, and indicates clearly that the primal plant is not a concrete principle but supersensible and formal. But he also says that this supersensible principle is present as form in the sensible matter of the plant. In addition, Goethe reveals that the original identity of all plant parts in the leaf is now completely transparent to him.

1.2 What is Goethe's primal plant and its organ, the leaf?

I will now introduce a variety of readings of the primal plant and its organ, the leaf, that have been suggested by scholars. While they rarely interpret the primal plant and the leaf as material principles of plant generation and development (Dornelas and Dornelas 2005, partly Brady 1987), some scholars think that they are formal principles (partly Brady 1987, Arber 1970, Schiller). A larger group of authors has suggested material formal readings (Steiner 1897/[5]1921, Benn 1932 in 1977, Heusser 2008, Wellmon 2010), and some scholars emphasize that the primal plant and the leaf are mutable, changeable principles (Bockemühl 1964, Nassar 2011, Förster 2011).

1.2.1 The primal plant and its organ, the leaf, as a material principle

To my knowledge, scholars rarely ascribe to Goethe a concrete, empirical notion of the primal plant and its organ, the leaf. But Marcelo C. Dornelas and Odair Dornelas (2005, 337), for instance, have defended what they claim is an updated, modern reading of Goethe's notion of the leaf, in which they interpret it as A-function "genes with homeotic effects" that initiate the metamorphosis of the leaf into other parts of the plant and those into further parts (I will return to details of this interpretation in Section 2.2.2).

Ronald H. Brady (1987) also mentions the possibility of this interpretation when he notes, correctly I guess, that some of Goethe's earlier notes on the primal plant seem to describe a primal concrete member, an ancestor of the plant genus or species, even though, notes Brady, Goethe develops a more abstract notion of it in his later notes. Brady writes:

> Goethe begins by speaking of his *Urpflanze* as if it were an ancestral form. His later references to it make it clear, however, that it is something abstracted from the empirical particulars.... [W]hen he speaks of discovering the *Urpflanze* "among this multitude" he probably does not mean actually *finding* it among the plants of the Palermo garden but *seeing* it through their mediation. After all, Goethe boasts, a month later, that "Nature herself" would envy him the *Urpflanze*, clearly indicating that it was not a natural product. Unless he went through a very rapid conversion, the *Urpflanze* was probably a general plan – rather than an ancestral species – from his inception.
>
> (Brady 1987, 268–269)

1.2.2 The primal plant and its organ, the leaf, as a formal principle

Agnes Arber (1970) also rejects the notion of the primal plant as a concrete ancestor in her reflections on English translations of Goethe's notion *Urpflanze*:

> If we translate *Urpflanze* as "primitive plant," or "primaeval plant," we are reading into it an evolutionary meaning which would have been foreign to Goethe's mind. To him the *Urpflanze* was a concept, from which the concepts of existing plant forms could be derived mentally; it carried no phylogenetic implications, and did not to him suggest any notion of an ancestral stock.
>
> (Arber 1970, 59; see also 42–45)

Instead of a concrete, empirical reading, even in view of Goethe's earliest notes on the primal plant and its organ, the leaf, Arber's reading takes a rationalist turn. She considers the primal plant an eternal idea with some resemblance to Platonic forms from which the forms of existing plants can be derived when they participate in the nature of the primal plant (Arber 1970, 67). Arber denies that the primal plant once existed in the past and emphasizes the non-historic character of this notion.

Arber is also one of the few interpreters who is sensitive to the various extensions of the "leaf" or type (*Pflanzentypus*) concept. She thinks that within the plant itself Goethe postulated "a single type for different lateral appendages of the stem (e.g. foliage-leaves and flower-parts) and other types of higher orders for different kind of shoots (e.g. leavy shoots and flowers; or expanded inflorescences and capitula)"; in the plant as a whole Goethe "hinted at the concept of a type for each family of plants"; and, finally, "Goethe postulated an archetype for all flowering plants – the *Urpflanze*" (Arber 1970, 61). Thus, the primal plant and its organ, the leaf, represents the entire genus of plants. Interestingly, Arber remarks that, from a (modern) botanist's perspective, the primal plant could be "bound up with the idea that the phyllome is an organ *sui generis*" (Arber 1970, 58), since the phyllome is a general plant pattern: it can be a foliar plant, or any organ homologous with a leaf, or a structure corresponding to a leaf, or produced by metamorphosis of a leaf.

Another rationalist reading of Goethe's notion of the primal plant and its organ, the leaf, stems from one of Goethe's most famous traditional interpreters: Friedrich Schiller. Goethe remembers one of his signature conversations with Schiller in his autobiographical notes *Glückliches Ereignis* (1817a).[2] He retells the moment in which he conversed with Schiller about his detection of the primal, "symbolical plant (*symbolische Pflanze*)" and its metamorphosis. Schiller, who listened carefully to Goethe's description, replied: "This is no experience, this is an idea" (HA 10.540), to which Goethe, disappointedly, responded: 'This can be very dear to me, that I have ideas without knowing it, and even see them with eyes' (HA 10.541). In this conversation, Schiller, a Kantian at this time, understood Goethe's notion of the primal plant and its organ, the leaf, as a Kantian regulative idea, that is, as a concept of reason that represents the necessary unity of the empirical manifoldness of nature, but that, as an idea of reason, cannot be represented in experience. As a Kantian idea, it is also not generative for empirical objects. Goethe felt clearly misunderstood, since he actually *saw* the primal plant in the concrete nature around him.

1.2.3 The primal plant and its organ, the leaf, as a material formal principle

Another type of interpretation, which I will label here "synthetic" or material formal reading, can be found, for instance, in Steiner (1897/[5]1921), Benn (1932 in 1977), Heusser (2008), and Wellmon (2010).

Rudolf Steiner (1897/[5]1921) in his book *Goethes Weltanschauung* was one of the earliest interpreters of Goethe's biological writings who developed a synthetic reading of the notion of the primal plant and its organ, the leaf, although some of Steiner's remarks could sound like a defense of a rationalist reading, when he writes: "Goethe thinks that the entire idea (*Idee*) of the realm of plants is contained in the primal plant, and the entire idea (*Idee*) of the realm of animals in the primal animal" (Steiner 1897/[5]1921, 109), or says that:

> Goethe has searched for the ideas of the primal plant and the primal animal in order to find in them the explanatory grounds of the manifoldness of organic forms. The primal plant is the generative element in the realm of plants. In order to explain a single kind of plant one has to demonstrate how the generative element is productive in the particular case.
>
> (Steiner 1897/[5]1921, 110)

But Steiner's more complex synthetic, material formal reading of Goethe's notion of the primal plant and its organ, the leaf, comes to the fore in the context of his interpretation of Goethe's notion of the metamorphosis of plants. There Steiner writes that Goethe believes:

> that ideas as much belong objectively to the objects as that what we can perceive with the senses. Goethe thinks that one can speak of the transformation

> of an organ into another only if … both contain something which is common to both. This is the sensible-supersensible form. The stamen of a plant form can only then be considered the transformed leaf of its ancestors, if the same sensible-supersensible form is alive in both. If this is not the case, then a stamen develops at the same place of the plant form at which a leaf had developed in its ancestors, and then nothing has undergone a transformation, but only one organ has replaced another.
>
> (Steiner 1897/[5]1921, 115)

Though speaking of a "sensible-supersensible form" is contradictory in a strict sense, what Steiner probably has in mind is that Goethe's idea of the leaf, the organ of the primal plant, is an idea that is in the object. It is the appearance of the supersensible form of the leaf in a sensible, concrete part of the plant.

More directly in traditional philosophical terms, Peter Heusser (2008) emphasizes that Goethe's understanding of natural science is a synthetic "rational empiricism (*rationelle Empirie*)" and "realism of ideas (*Ideenrealismus*)" (Heusser 2008, 111). Heuser sees Goethe among those scientists who consider an idea a law of nature that "is not only a subjective principle of order, but also an entity that is objectively active in nature" (Heusser 2008, 113). He thinks that Goethe avoids one-sided empirical and rational interpretations of the leaf, since, according to Goethe, "laws (ideas), are objective, immanent principles of nature, which can be accessed by the mind" (Heusser 2008, 114).

Also Gottfried Benn (1932 in 1977, 175–177), in equally strong terms, though those of a powerful poet, offers a synthetic reading of Goethe's primal plant and its organ, the leaf. He understands Goethe's conception as an "idea," but a very special one – an idea that is "an identity, that moved, a reality, that became dialectical, a worldliness in which transcendence activated itself." Benn writes that Goethe's idea of the primal plant was "a concept that mediated between the lawfulness of eternal forms and the creative freedom of life"; it was "the whole in the universal" and "always the same monistic totality, out of which the particular becomes concrete." And Benn also points to the double nature of the primal plant and its organ, the leaf, when he calls it a totality and a particular.

Finally, Chad Wellmon (2010) holds a synthetic reading with a historical twist, when he claims that:

> Linnaeus's taxonomic tables were merely storage devices that subordinated … the study of the interrelation of natural forces and natural historical change … to the classification of nature according to taxonomic schemes. Linnaeus simply catalogued nature. Goethe credits Jean-Jacques Rousseau, however, with not only having introduced him to the broader discipline of botany but having promised "a method less opposed to the senses."
>
> (Wellmon 2010, 156)

1.2.4 The primal plant and its organ, the leaf, as a mutable, dynamical principle

A further interesting reading of the primal plant, and its organ, the leaf, which is different from the former interpretations, has been suggested by Förster (2011) and Nassar (2011). Dalia Nassar notes:

> What grants the plant unity is not a static substance or idea, but the fact that the plant is in a process of metamorphosis, wherein each part is a physical manifestation of the different stages of the metamorphosis. Goethe is thus able to determine the unifying principle that underlies all plants, the *Urpflanze*. The *Urpflanze* is not an abstract, subjectively determined principle – i.e., it is not a postulate of thought. Rather, it is the living principle that underlies and organizes every plant. It is the principle of unity that, as Goethe puts it, can only be gleaned amid multiplicity.
>
> (Nassar 2011, 75)

Nassar, if I take her right, describes Goethe's primal plant and its organ, the leaf, as a dynamical ("living"), objective, and concrete ("not an abstract, subjectively determined") principle, as an "underlying productivity" (Nassar 2011, 75). It is neither eternal and unchangeable as a Platonic idea of the genus or species of plants nor an unchangeable Kantian postulate, a regulative idea of the unity of the empirical manifoldness of plants. Instead, according to Nassar, Goethe's primal plant is a dynamical, changeable, objective principle that underlies all plant generation and development.

Since Nassar rejects the formal, conceptual nature of that primal plant and its organ, the leaf, what she has in mind is perhaps a material principle of the Presocratic kind, something that Thales imagined when he claimed that water is the principle of all things since it can transform into solid, fluid and gaseous states, the basic aggregate states of all concrete things. Or what Anaximenes wanted when he said that air is the principle of all things because it can transform into all particular things by means of expansion and compression. For Thales and Anaximenes, water and air are no static principles of being but are dynamical and changeable themselves. Nassar's reading captures better than any other reading what Goethe meant when he named the primal plant and its organ, the leaf, a mutable, changeable Proteus, though Nassar's reading fails to relate to the conceptual nature of the primal plant, and its organ, the leaf, that Goethe frequently emphasizes.[3]

Another version of a dynamical principle has been suggested by Eckart Förster (2011). Förster notes:

> But what exactly is it that is doing the expanding and contracting in these six stages [of the metamorphosis of plants]? It is obviously not any one of the visible parts of the plant. The sepal, for example, does not emerge from the stem leaf in any physical sense; it just follows upon it. So what expands and contracts and becomes concrete … in the individual parts is, in the first

> instance, a form that can only be apprehended in thought. When Goethe first realized this he made the following note in his diary: "Hypothesis: Everything is leaf, and the greatest diversity becomes possible through this simplicity" (LA II,9A:55). What he means by "leaf" is not the sensuously given stem leaf, but rather an ideal organ from which all the physical forms of the plant can be formed by way of transformation so that the petals, too, and the stamen and pistils must be considered as metamorphosed leaves. The fact that they look different when viewed superficially is itself a merely superficial fact. According to the hypothesis, the plant's visible parts are merely particular formations of an underlying ideal form which presents itself anew at every nodal point, repeating its work. Thus, "everything is leaf" must be understood as meaning that all the forms taken on by the plant are, *in respect to their idea*, identical.
>
> (Förster 2011, 274; 2012, 273–274)

According to Förster, the primal plant and its organ, the leaf, is no material or physical principle but the "ideal form" of all plants and their parts (*eine zunächst nur gedanklich fassbare Form*), which is active in plant matter. It is a mutable principle that transforms into the manifoldness of concrete parts of the plants by means of expansion and contraction.

1.2.5 Which reading is right?

What kind of principle is the primal plant, and its organ, the leaf? Given the textual ambiguity in Goethe's notes it is hard to decide which of the readings is right, since the various interpretations are dependent upon the emphasis that the different interpreters put on particular aspects of Goethe's notes. But in my view there are a couple of thoughts that Goethe repeats more frequently than others.

Goethe consistently claims that he is becoming aware of the idea or concept of the primal plant and its organ, the leaf, when and while observing the surrounding manifoldness of plants in Italy. Goethe's primal plant and its organ, the leaf, is present in nature and cannot be "just" an idea, a thought, separate from matter and the physical world. Thus I disagree with Arber's (1970, 59) and Schiller's (HA 10.540–10.541) readings. What is false in Schiller's Kantian take of Goethe's idea is that the Kantian idea can never be fully represented in nature (experience), but Goethe claims to *see* and *observe* the primal plant and its organ, the leaf, in nature. A Kantian idea is also not generative, but Goethe thinks of the primal plant and its organ, the leaf, as enabling not only the human recognition of all plants but also plant generation and growth. In Arber's reading Goethe's idea of a primal plant, and its organ, the leaf, seems to be similar to a Platonic universal that ontologically and epistemologically enables the existence and the recognition of concrete plants as corresponding particulars. But, though Platonic ideas have a greater fit with Goethe's idea of a primal plant and its organ, the leaf, than Kantian ideas, since they are generative, the disadvantage of

Platonic ideas is that they are not in nature but are separate, supersensible entities. But Goethe claims to see the idea of the primal plant and its organ, the leaf, in nature. And both Kantian and Platonic ideas do not fulfill the mutability requirement of Goethe's principle.

Among the synthetic readings, Steiner (1897/[5]1921, 115) seems to consider Goethe's notion of the primal plant and its organ, the leaf, as a sensible-supersensible form of an ultimate ancestor. Steiner is also more aware of the generative force of the primal plant, and its organ, the leaf, than Schiller (Steiner 1897/[5]1921, 110). But it is also another thing to say that the idea is in nature, as form is in matter, than that the idea itself is sensible-supersensible, which is contradictory. And it is perhaps not necessary to locate the idea of the primal plant and its organ, the leaf, as an ancestor in the past.

Nassar's (2011, 75) description of Goethe's primal plant as a dynamical ("living"), objective, and concrete ("not an abstract, subjectively determined") principle captures best the mutable and generative aspect of Goethe's notion but fails to incorporate Goethe's frequent description of the primal plant and its organ, the leaf, as form, concept, or idea. And this is the advantage of Förster's (2011, 274; 2012, 273–274) reading, in that it defends a mutable and at the same time formal principle, which is active in matter. Although it is not entirely clear whether Förster considers the leaf the idea of the plant genus (*the* plant) or a variety of ideas of particular plant species, Förster's reading seems to be closest to Goethe's notes.

2 The metamorphosis of the leaf

So far we have considered possible readings of Goethe's notions of the primal plant and its organ, the leaf, as the principle of plant generation and development. I will now consider how the plant generation and development proceeds from the primal plant and its organ, the leaf. In order to do so, I will first consider Goethe's description of the progressive metamorphosis of plants (Section 2.1). Then I will discuss possible readings of this metamorphosis, especially in view of the generic nature of the primal plant and its organ, the leaf (Section 2.2). I will discuss three models of plant generation and growth (Sections 2.2.1–2.2.3) and will consider which of them allows for a meaningful interpretation of Goethe's mysterious claim that "all is leaf" and at the same time is coherent with our common experience of plant generation and development (Section 2.2.4).

2.1 Goethe's theory of the metamorphosis of the leaf in the **Metamorphosis of Plants**

In Sicily Goethe detects the "*original identity* of all parts of a plant," namely that each "sensual form" of the plant expresses the same "supersensible primal plant" (HA 13.164). In the *Metamorphosis of Plants* Goethe identifies this identical supersensible principle in each sensual form as the leaf.[4] Goethe is convinced then that all parts of a plant develop out of one and the same organ, the leaf:[5]

> Whenever then the plants vegetate, blossom, or bear fruits, it nevertheless is always the same organs, with varying functions and with frequent changes in form, that fulfill the dictates of Nature. The same organ which expanded on the stem as a leaf and assumed a highly diverse form, will contract in the calyx, expand again in the petal, contract in the reproductive organs, and expand for the last time as fruit.
>
> (*Met.* 115, HA 13.100)

The process of the transformation of the primal plant into a concrete plant proceeds through the organ, the leaf. Goethe names this process, as did some of his predecessors, "metamorphosis":

> The intimate relationship of various external plant parts – such as leaves, calyx, corolla, and stamens – which develop one after another, and apparently from another, has long been recognized by naturalists ... and the process by which one and the same organ makes its appearance in multifarious forms has been named the *metamorphosis of plants*.
>
> (*Met.* 4, HA 13.64)

He distinguishes between three kinds of metamorphosis: a regular, an irregular, and an accidental one (*Met.* 5, HA 13.64). Whereas the regular or progressive metamorphosis concerns the plant transformation from the first seed leaves to the final development of the fruit, the irregular or regressive metamorphosis is a transformation, in which the plant slackens and decays and is in an inwardly feeble and ineffective state (*Met.* 7, HA 13.65). The accidental metamorphosis is effected by outside agents, especially insects (*Met.* 8, HA 13.65).

Goethe claims that the regular or progressive metamorphosis proceeds in "six steps" (*Met.* 73, HA 13.85–13.86; see also *Met.* 115, HA 13.100) by means of the organ, the leaf, and two forces, expansion and contraction:

1 The seed (*Samen*) develops into seed leaves (cotyledons, *Samenblätter*, *Kotyledonen*) and stem leaves (*Stengelblätter*) through an expansion of the leaf/leaves (*Met.* 10–18, HA 13.66–13.69; see lines 11–32, HA 1.199[6]).

2 The stem leaves (*Stengelblätter*) metamorphose into sepal leaves (*Kelchblätter*) and the calyx (*Kelch*) through a contraction of the leaf/leaves (*Met.* 19–28, HA 13.69–13.72; see lines 33–43, HA 1.200[7]).

3 The sepal leaves (*Kelchblätter*) and the calyx (*Kelch*) develop into petal leaves (*Blütenblätter*) and the corona (*Krone*) through an expansion of the leaf/leaves (*Met.* 29–45, HA 13.72–13.77; see lines 44–50, HA 1.200[8]).

4 The petal leaves (*Blütenblätter*) and the corona (*Krone*) transform into the staminal organs (*Staubwerkzeuge*) and the style (*Griffel*) through a contraction of the leaf/leaves (*Met.* 46–73, HA 13.77–13.86; see lines 51–56, HA 1.200[9]).

5 The staminal organs (*Staubwerkzeuge*) and the style (*Griffel*) develop into the fruit (*Frucht*) through an expansion of the leaf/leaves (*Met.* 74–81, HA 13.86–13.87; see lines 57–59, HA 1.200[10]).

6 Finally, the fruit (*Frucht*) transforms into the new seed (*Samen*) through a contraction of the leaf/leaves (*Met.* 82–83, HA 13.88–13.89; see lines 60–62, HA 1.200[11]).

As evidence for the six steps of the progressive metamorphosis of the plant, Goethe also points to the six corresponding reversed processes in the retrogressive metamorphosis: as much as a sepal leaf and calyx can be considered an expanded stem leaf (progressive), a stem leaf can be considered a contracted sepal leaf and calyx (retrogressive); as much as stamina and style can be considered contracted petal leaves and a corona (progressive), the petal leaves and the corona can be considered expanded stamina and a style (retrogressive), etc.

Goethe thinks that the processes of expansion and contraction in each step of the metamorphosis are caused by a living force (*Lebenskraft*, *Met.* 113, HA 13.99). This living force expresses itself successively and simultaneously: successively, because the development of the petal leaves and the corona follows the development of the stem and the stem leaves, and the development of stem leaves follow that of the seed leaves and seed; simultaneously, since all existing parts of the plant grow at the same time. Though this outline of Goethe's account sounds clear in general, it is tricky and unclear upon closer consideration. How exactly does the metamorphosis of plants proceed? Inasmuch is the principle of plant generation and development, the leaf, involved? Again, scholars have suggested different readings of these issues, which I would like to compare and discuss now.

2.2 How does the regular metamorphosis of plants proceed and which role does the leaf play in it?

In order to answer the questions how the regular metamorphosis of plants proceeds and which role the leaf as the principle of plant generation and development plays in it, I will now distinguish and discuss three models. Suppose the leaf is A and the parts of the plants are B, C, and D. Does A transform into B, C, D by a process from A to B to A, from A to C to A, and from A to D to A (Section 2.2.1)? Or does A transform into B, proceed from B as B to C, and proceed from C as C to D (Section 2.2.2)? Or does A transform into B, while remaining at the same time A, and proceed from B as A to C, while remaining at the same time A, and proceed from C as A to D, while remaining at the same time A (Section 2.2.3)? I will discuss whether these three models allow a meaningful interpretation of Goethe's mysterious claim that "all is leaf" and whether they are coherent with the common experience of plant generation and development. I will argue that model 1 accounts primarily for a plausible explanation of the permanence of the principle of plant generation and development, the leaf, but does so at the cost of a plausible explanation of the resulting metamorphosis. Model 2 primarily accounts for a plausible explanation of the metamorphosis of the plant, but does so at the cost of an explanation of the permanence of its principle, the leaf. Model 3 is the most promising reading, but there are difficulties remaining (Section 2.2.4).

2.2.1 The metamorphosis of the plants and the leaf: model 1

Suppose the leaf is A and the parts of the plants are B, C, and D. Does A transform into B, C, and D by a process from A to B to A, from A to C to A, and from A to D to A? A first reading of Goethe's notion of a metamorphosing plant development could be that the plant is generated by its principle, the leaf (A), and then develops progressively, first into seed leaves and stem leaves (B), then returns to principle A in order to develop into C (the calyx and sepal leaves), and then returns to principle A in order to develop into D (the corona and petal leaves), etc.

Model 1 nicely represents Goethe's idea that the leaf (A) is the principle of every step of plant generation and development by means of a repeated return to A after each step of the plant development. But it is difficult to see how model 1 captures Goethe's idea that in our observation of plant growth the parts of the plant seem to develop successively one after another: the seed develops, followed by the stem and its stem leaves, the stem and its stem leaves followed by the calyx and its sepal leaves, the calyx and its sepal leaves followed by the corona and its petal leaves, the corona and its petal leaves followed by the stamen and the style, and those by a fruit and seed.

If we would understand the repeated return to the leaf (A) as the principle in a temporal sense, each progressive phase of plant growth would have to be followed by the same regressive phase. The seed, for instance, after having developed into a stem and stem leaves, would have to decay as stem and stem leaves, in order to make room for the development of a new part of the plant, for instance the calyx and sepal leaves. But this is not in harmony with the common observation of plant development. If we, in turn, would understand the return to A as a spatial one, we would observe a zig-zag ontology, in which one part of the plant develops, followed by another one, which begins at the same spatial point from which the former part began to develop and not at its end point. But also such a spatial return to the leaf as the principle of plant development is not in harmony with the common observation of plant development.

2.2.2 The metamorphosis of plants and the leaf: model 2

Let us consider a second model. Suppose the leaf is A and the parts of the plants are B, C, and D. Does A transform into B, proceed from B as B to C, and proceed from C as C to D? Model 2 suggests that A is the principle at the beginning of the plant generation and development, and everything in the plant is leaf in the sense that it is dependent upon the leaf as the ultimate starting point of plant generation and development. The subsequent steps of plant growth are no longer directly caused by the leaf but are transformations of one part of the plant into another part.

Several of Goethe's remarks seem to support such a view. At the beginning of the *Metamorphosis of Plants* Goethe notes: "Anyone who devotes the least attention to the growth of plants can easily note that certain of their external parts are often transformed, assuming, either completely or to some lesser degree, the form of the neighboring parts" (*Met.* 1, HA 13.64). In *Met.* 1 (HA 13.64), Goethe seems to suggest that the subsequently developing parts do not necessarily resemble the form of the primal leaf but rather the immediately preceding and following parts. Similarly, he says:

> Nature's regular procedure and … the laws of transformation … bring forth one part through another, achieving the most diversified forms through modification of a single organ.
>
> (*Met.* 3, HA 13.64)

> The intimate relationship of various external plant parts … which develop one after another, and apparently from one another, has long been recognized by natural scientists.… [A]nd the process by which one and the same organ makes its appearance in multifarious forms has been named the *metamorphosis of plants*.
>
> (*Met.* 4, HA 13.64)

Both statements are slightly ambiguous with regard to the role of the primal organ of the plant, the leaf. When Goethe insists that the external parts of the plant seem to develop "through another" (*Met.* 3, HA 13.64) or "apparently from one another" (*Met.* 4, HA 13.64) he seems to suggest that neighboring parts of a plant develop from one another, but not (at least not directly) from one and the same primal organ of the plant, the leaf. Observation reveals a "transmutation of one form into another" (*Met.* 6, HA 13.64–13.65) but not a transmutation of every external part of the plant from one and the same primal organ of the plant, the leaf. Nevertheless, Goethe also emphasizes exactly this, namely that all external parts of the plant develop from one and the same primal organ of the plant, the leaf.

What speaks against model 2 is, thus, that it does not capture well enough the permanent presence of the same principle, the leaf, in all steps of plant development. In model 2, the later developing parts of the plant lose more and more contact to the principle of plant development. The leaf only initiates the beginning, but does not renew its causal influence on the process of plant development. So one could conjecture that, for instance, the corona and petal leaves, the stamina and style, and finally the plant's fruit, no longer show strong influences of the primal organ, the leaf. But the corona and the petal leaves, stamina, style, and finally the plant's fruit are certainly identifying parts of a plant – even more identifying ones than earlier parts like the seed and stem leaves.

The biologists Enrico S. Coen (2001) and Marcelo C. Dornelas and Odair Dornelas (2005, 337–338) offer an interpretation of Goethe's leaf archetype and its transformations, which is in some way close to model 2. They argue from the perspective of contemporary evolutionary developmental biology and apply what they call the "ABC model of flower organ identity" to Goethe's theory.

The ABC model of flower development or flower organ identity was originally developed by Enrico S. Coen and Elliot M. Meyerowitz (1991) through the study of homeotic mutations in *Arabidopsis thaliana*, the thale cress, mouse-ear cress, or *arabidopsis*. As a developing flower this plant has four basic organs: sepals, petals, stamens, and carpels (which go on to form pistils), which are arranged in a series of whorls: four sepals on the outer whorl, followed by four petals inside this, six stamens, and a central carpel region. The homeotic mutations of these organs by homeotic genes, that is, genes that regulate the development of anatomical structures in organisms through the programming of transcription factors, result in the change of one organ to another – for example, stamens become petals, and carpels are replaced with a new flower, resulting in a recursively repeated sepal-petal-petal pattern.

The ABC model of flower development formalizes and generalizes the study of this growth. The model suggests that there are floral organ identity genes, which are divided into three classes: class A genes, which affect sepals and petals, class B genes, which affect petals and stamens, and class C genes, which affect stamens and carpels. Owing to an overlap of genes, the organs or parts of the plant can transform and metamorphose into each other. See Figure 8.1.

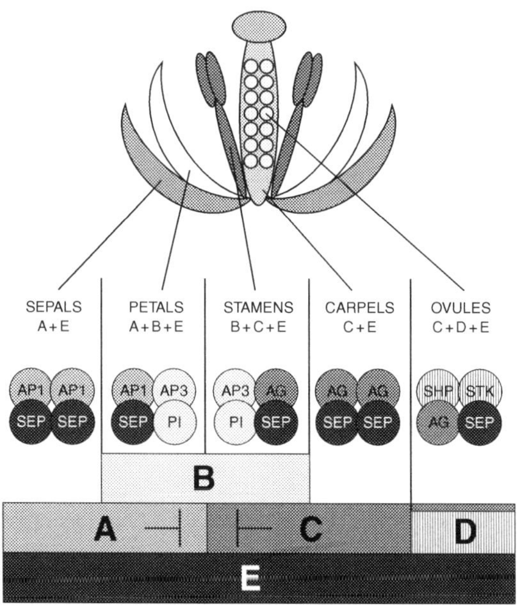

Figure 8.1 Dornelas and Dornelas's (2005, 338) revised ABC model of flower organ identity. Dornelas and Dornelas write: "A-function genes … are necessary for the formation of the sepals, B-function genes … together with A-function genes, are necessary for the formation of the petals. The B-function genes, along with those of C-function, are necessary for the formation of the carpels. The ABC model has been gradually expanded to include class D- and E-function genes, which are necessary for the ovules and the definition of the floral whorls, respectively."

Dornelas and Dornelas (2005, 337) claim that the ABC model "updates Goethe's theory that all parts regulated by ABC series conform to a generalized 'leaf' archetype." This is true to some extent, but to some extent not. What speaks against their reading? First, for Goethe the development and growth of the plant in the six metamorphoses of the leaf begins with a leaf that is active in the seed already and not upon the emergence of sepal or petal leaves. Second, the ABC model primarily explains the metamorphosing organs but not the presence of the same principle, the leaf, in all steps of this metamorphosis. But it should explain the latter as well. Third, the ABC model thus is confronted with the same objection that I mentioned against model 2 in general, namely that the distance of the newly developing organs and parts of the plant to the principle of plant development and growth, the leaf, is getting larger with each transformation of one organ into another. But the parts in further distance to the principle of plant generation and development might have the same or an even larger identifying potential for a specific plant than parts which are close to this principle. And, fourth, the ABC model of flower identity interprets the principle of plant development, the leaf, as a genetic starting code. Though this is not as robust as sheer matter, it is a materialist interpretation of the leaf. And this contradicts Goethe's claim that the primal plant and its principle, the leaf, is a supersensible form that can be observed in sensible matter.

2.2.3 The metamorphosis of plants and the leaf: model 3

Let us consider a third and final model. Suppose the leaf is A and the parts of the plants are B, C, and D. Does A transform into B, while remaining at the same time A, and proceed from B as A to C, while remaining at the same time A, and proceed from C as A to D, while remaining at the same time A?

Let us have a second look at the passages (*Met.* 3–4, HA 13.64, my emphasis) in Goethe's *The Metamorphosis of Plants* that we have considered already, now with a specific emphasis:

> Nature's regular procedure and … the laws of transformation … bring forth one part through another, *achieving the most diversified forms through modification of a single organ.*
>
> The intimate relationship of various external plant parts … which develop one after another, and apparently from one another, has long been recognized by natural scientists.… [A]nd the process *by which one and the same organ makes its appearance in multifarious forms* has been named the *metamorphosis of plants.*

In these passages, Goethe is eager to emphasize both: that the parts of the plant develop after and from one another, and that in all steps of this development the leaf, the principle of plant generation and development, is equally involved and present. This can be approved by another passage, in which Goethe writes that "the leaves of the calyx *are the very same organs* which have previously made their appearance as stem leaves, but are now clustered around a common center, often in greatly altered form" (*Met.* 32, HA 13.73, my emphasis; see also *Met.* 50, HA 13.79; *Met.* 71, HA 13.59–13.60). While model 2 violated the formal character and the sameness of the leaf, the principle of plant development, in all its stages, model 3 not only offers an explanation of the metamorphosing transformation of the organs or parts of the plant, but also of the sameness of the principle of this metamorphosis, the leaf.

A first interpretation that represents model 3 can be found in Rudolf Steiner's (1897/[5]1921) text *Goethes Weltanschauung*. Goethe believes, Steiner notes,

> that ideas as much objectively belong to the objects as what one can perceive with the senses. Goethe thinks that one can speak of the transformation of an organ into another only if ... both contain something which is common to both. This is the sensible-supersensible form. The stamen of a plant form can only then be considered the transformed leaf of its ancestors, if the same sensible-supersensible form is alive in both.
>
> (Steiner 1897/[5]1921, 115)

According to Steiner, the primal plant and its organ, the leaf, is the generative force behind all plant generation and development. In order to account for this, Steiner characterizes the leaf as a sensible-supersensible principle. This slightly overstates Goethe's theory since Goethe describes the leaf as a formal principle that is present in matter, but not as a sensible-supersensible, material formal principle itself. In this way Steiner can avoid the objection that it is difficult to see how the permanently metamorphosing form(s) of the plant organs, which are caused by the leaf, can be realized in a suitable, equally mutable matter, since the principle of change itself is intrinsically connected to such matter. But Steiner simply assumes that there is such matter, and his argument seems to be (if there is one) that the relevant fit of form and matter once existed in an original ancestor of the plant in the past.

Let us consider a second variant of model 3. Jochen Bockemühl (1964, 1966, 1970), in several similar papers and accompanying drawings, suggests a reading of Goethe's metamorphosis of plants that supports model 3, since he gives numerous examples that demonstrate plausibly as to how the various parts of the plant transform into one another through transitions of their forms, which are caused by the generative principle of plant development, the leaf. Bockemühl (1964 in 1982, 10) describes the leaf as "the conceptual element of the plant type."

In the earliest of these papers, Bockemühl (1964 in 1982, 10) writes that a scientist who wants to recognize the laws of the plant organs and their spatial and temporal order, their growth and development, must recognize the conceptual element (*das begriffliche Element*), the type (*Pflanzentypus*) in it, which, he says, is not a pale universal or static blueprint but is in its own nature a changeable and mutable principle. See Figure 8.2.

Figure 8.2 These are two of the typical drawings with which Bockemühl illustrates his interpretation of Goethe's notion of a progressive metamorphosis of plant organs. The first (Brockemühl 1964 in 1982, 10) shows the complete series of foliage leaves of a sow-thistle (*Sonchus oleaceus L., Compositae*). The second (Bockemühl 1970 in 1982, 127) shows the complete series of leaves of a blooming shoot of the wild peony (*Paeonia broteroi*).

In order to demonstrate how this mutable principle effects plant development, Bockemühl (1964 in 1982, 10–12) asks us to perform a preparatory thought experiment with a geometrical example. We shall imagine the transformation of one triangle into another one by a stepwise change of the triangle's ankles and side lengths in accordance with certain rules. Bockemühl claims that the growth and development of plant organs can be imagined in a similar way, since we can derive the forms of plant organs from each other in the same way in which we can derive one form of a triangle from another one. The difference between the example of the triangles and plant organs is that in the case of triangles we ourselves choose the rules and direction of the change of the triangles' forms, whereas in the case of plant development we cannot choose or influence these rules or directions, but can only follow them, since nature, and more specifically the plant type, dictates these rules and directions of change.

Bockemühl remarks that, if we reconstruct the transformation of plant organs into each other in our thought, we understand the motions and transformations that seem to take place in the actual plant. But he thinks that what we understand then is only what seems to take place in the plant, since what we actually observe are not the transformations of the parts of the plant into each other but their already-completed organs. The motion and transformation that we recognize in our imagination and thought represents, for Bockemühl, the ideal relationship of plant organs to each other and their transformation into each other. Bockemühl thinks that it is this ideal relationship that Goethe means when he speaks about the progressive metamorphosis and the related processes of contraction and expansion which are caused by the principle of plant development, the leaf.

Finally, let us consider a third interpretation that supports model 3. For this interpretation we return to the passage by Eckart Förster (2011) that we considered already in our earlier discussion of the nature of the primal plant and its organ, the leaf. In his interpretation of Goethe's notion of the metamorphosis of plants and its relationship to the leaf, Förster writes that it is "not any one of the visible parts of the plant" that is "doing the expanding and contracting" in the six metamorphoses of the leaf, but "a form that can only be apprehended in thought," the leaf. The leaf is an "ideal organ from which all the physical forms of the plant can be formed by way of transformation." According to Förster's interpretation "the plant's visible parts are merely particular formations of an underlying ideal form which presents itself anew at every nodal point, repeating its work." All the forms taken on by the plant "are, *in respect to their idea*, identical" (Förster 2011, 274; 2012, 273–274). According to Förster, the primal plant and its organ, the leaf, is the form, which expands and contracts in the six steps of the metamorphosis of the plant. It is at each step of the plant generation and development present as the same mutable formal principle that initiates the plant's growth through expansion and contraction.

2.2.4 Which of the three models is right?

Whereas model 1 accounts for a plausible explanation of the permanence of the principle of plant generation and development at the cost of a plausible explanation of the resulting metamorphosis, model 2 accounts for a plausible explanation of metamorphosis of the plant at the cost of an explanation of its permanent principle. Among the three versions of model 3, which is the best interpretation available so far, Steiner's (1897/[5]1921) reading has the problem that it departs from Goethe's text, when it describes the principle of plant generation and growth as a sensible-supersensible, material formal principle, whereas Goethe just considers it a form that is active in matter. Bockemühl (1964, 1966, 1970) and Förster (2011) stay closer to Goethe's text and interpret the leaf as a formal principle. Bockemühl draws more attention to a plausible reconstruction of the metamorphosing parts of the plant, but does not forget to tell us that they are an effect of the conceptual element, the plant type or leaf. Förster's view is more balanced since Förster shows that the leaf as the living force causes expansion and contraction, and demonstrates then that expansion and contraction are the formative forces behind the formation of every concrete part of the plant. A remaining problem for Bockemühl's and Förster's readings is how to explain how the formal principle of plant generation and growth, the leaf, which causes the various metamorphoses of the parts of the plant, fits into the plant's matter. Here one would wish to have an illuminating eighteenth-century theory of plant matter – something similar to the modern theory of plant matter that Dornelas and Dornelas (2005) describe, or to the phyllome that Arber (1970, 58) mentions.

Such a specific theory of matter would be required, since not every matter is equally suitable for the leaf's forming processes of expansion and contraction, and not every matter is capable of the constitution of a living being.[12]

Conclusion

Though Goethe developed his theory of plant morphology and metamorphosis without many modern insights, which shape our botanical knowledge today – biogeography and ecology, cell theory, molecular biology, plant biochemistry, and economic botany – his metamorphosis of plants remains inspiring as much for Goethe scholars and natural philosophers, as for biologists, especially botanists. And this surely has something to do with Goethe's fundamental questions concerning principles of plant generation and development, the matter and form of plants, and the life cycle of plants in general, questions that require and challenge philosophical answers as well as those of the botanical sciences of the day.

Abbreviations

HA *Goethes Werke. Hamburger Ausgabe* (14 vols.), Munich: C. H. Beck 1999.
Met. Goethe, J. W. 1790. Die Metamorphose der Pflanzen, in *Goethes Werke. Hamburger Ausgabe. Naturwissenschaftliche Schriften I* (vol. 13, 64–101), Munich: C. H. Beck 1999.

Notes

1 Goethe's notion of a 'höhere Anschauung' and its relation to Kant's notion of an intellectual intuition has received much attention during the last years the discussion of which I have to postpone to another occasion. For relevant literature see, for instance, Förster's work, Bowman (2011), and Hindrichs (2011).
2 As is well-known, this is one of the most commented autobiographical events in the development of Goethe's views on plant metamorphosis; see, for instance, Steiner (1897/[5]1921, 17–20), Arber (1970, 61–62), Förster (2001, 90), and Steigerwald (2002, 308).
3 Nassar (2011, 76–87) shows that the recognition of the metamorphosis of the plant requires the intuitive understanding of nature's observer, but fails to see the conceptual nature of the leaf and its productivity itself. The conceptual element that Goethe claims does not appear when an observer of nature engages with the recognition of nature, but is a characteristic of living nature itself.
4 Some scholars note (Arber 1970, 33–58; Brady 1987, 269) that Goethe was not the first who developed the idea of the fundamental identity of all plant organs as leaves. Variants of this idea appeared in Nehemiah Grew's (1641–1712) *The Anatomy of Vegetables Begun* (1672) and his *The Anatomy of Plants* (1682), in Marcello Malpighi's (1628–1694) *Anatome Plantarum* (1675/1679), and in Caspar Friedrich Wolff's (1734–1794) *De Formatione Intestinorum* (1768), though it is believed that Goethe was not acquainted with the works of these predecessors.
5 At several places Goethe also speaks of "organs" (plural); see *Met.* 115, HA 13.100.
6 In lines 11–32 (HA 1.199) of the poem "Die Metamorphose der Pflanzen" ("The Metamorphosis of Plants") Goethe describes the generation and development of the

first step of the metamorphosis of plants from the seed to seed and stem leaves through expansion of the leaf:

> Aus dem Samen entwickelt sich, sobald ihn der Erde
> Stille befruchtender Schoß hold in das Leben entläßt,
> Und dem Reize des Lichts, des heiligen, ewig bewegten,
> Gleich den zärtesten Bau keimender Blätter empfiehlt.
> Einfach schlief in dem Samen die Kraft; ein beginnendes Vorbild
> Lag, verschlossen in sich, unter die Hülle gebeugt,
> Blatt und Wurzel und Keim, nur halb geformet und farblos;
> Trocken erhält so der Kern ruhiges Leben bewahrt,
> Quillet strebend empor, sich milder Feuchte vertauend,
> Und erhebt sich sogleich aus der umgebenden Nacht.
> Aber einfach bleibt die Gestalt der ersten Erscheinung;
> Und so bezeichnet sich auch unter den Pflanzen das Kind.
> Gleich darauf ein folgender Trieb, sich erhebend, erneuet,
> Knoten auf Knoten getürmt, immer das erste Gebild.
> Zwar nicht immer das gleiche; denn mannigfaltig erzeugt sich,
> Ausgebildet, du siehst's, immer das folgende Blatt,
> Ausgedehnter, gekerbter, getrennter in Spitzen und Teile,
> Die verwachsen vorher ruhten im untern Organ.
> Und so erreicht es zuerst die höchst bestimmte Vollendung,
> Die bei manchem Geschlecht sich zum Erstaunen bewegt.
> Und viel gerippt und gezackt, auf mastig strotzender Fläche,
> Scheinet die Fülle des Triebs frei und unendlich zu sein.

7 In lines 33–43 (HA 1.200) of the poem Goethe describes the generation and development of the second step of the metamorphosis of plants from the stem leaves to the petal leaves and calyx through contraction:

> Doch hier hält die Natur, mit mächtigen Händen, die Bildung
> An und lenket sie sanft in das Vollkommnere hin.
> Mäßiger leitet sie nun den Saft, verengt die Gefäße,
> Und gleich zeigt die Gestalt zärtere Wirkungen an.
> Stille zieht sich der Trieb der strebenden Ränder zurücke,
> Und die Rippe des Stiels bildet sich völliger aus.
> Blattlos aber und schnell erhebt sich der zärtere Stengel,
> Und ein Wundergebild zieht den Betrachtenden an.
> Rings im Kreise stellt sich nun, gezählet und ohne
> Zahl, das kleinere Blatt neben dem ählichen hin.
> Um die Achse gedrängt, entscheidet der bergende Kelch sich…

8 In lines 44–50 (HA 13.200) Goethe describes the generation and development of the third step of the plant metamorphosis from sepal leaves and the calyx to the petal leaves and the corona through expansion:

> … Der zur höchsten Gestalt farbige Kronen entläßt.
> Also prangt die Natur in hoher, voller Erscheinung,
> Und sie zeiget, gereiht, Glieder an Glieder gestuft.
> Immer staunst du aufs neue, sobald sich am Stengel die Blume
> Über dem schlanken Gerüst wechselnder Blätter bewegt.
> Aber die Herrlichkeit wird des neuen Schaffens Verkündung;
> Ja, das farbige Blatt fühlet die göttliche Hand…

9 In lines 51–56 (HA 1.200) Goethe describes the generation and development of the petal leaves and corona into the staminal organs and the style through contraction:

Und zusammen zieht es sich schnell; die zärtesten Formen,
Zwiefach streben sie vor, sich zu vereinen bestimmt.
Traulich stehen sie nun, die holden Paare, beisammen,
Zahlreich ordnen sie sich um den geweihten Altar.
Hymen schwebet herbei, und herrliche Düfte, gewaltig,
Strömen süßen Geruch, alles belebend, umher.

10 In lines 57–59 (HA 1.200) Goethe describes the generation and development of the staminal organs and the style into the fruit through an expansion:

Nur vereinzelt schwellen sogleich unzählige Keime,
Hold in den Mutterschoß schwellender Früchte gehüllt.
Und hier schließt die Natur den Ring der ewigen Kräfte:

11 Finally, in lines 60–62 (HA 1.200) Goethe describes the transformation of the fruit into the new seed through contraction:

Doch ein neuer sogleich fasset den vorigen an,
Daß die Kette sich fort durch alle Zeiten verlänge
Und das Ganze belebt, so wie das Einzelne sei.

12 Reill (1986) points to the importance of an analysis of Goethe's notion of matter, which, in his view, is supposed to support Goethe's attempt to bridge the gap between empiricist and rationalist conceptions of the living. But, apart from some general hints on Goethe's notion of polarity (Reill 1986, 141–146), Reill does not engage in developing a detailed interpretation of Goethe's concept of plant matter.

References

Amrine, F., Zucker, F. J., and Wheeler, H. (eds.). 1987. *Goethe and the Sciences: A Reappraisal*, Boston, MA: Springer.

Arber, A. 1970. *The Natural Philosophy of Plant Form*, Darien, CT: Hafner.

Benn, G. 1932. Goethe und die Naturwissenschaften, in Wellershoff, D. (ed.). *Gesammelte Werke in vier Bänden* (vol. 1, pp. 162–200), Stuttgart: Klett-Cotta.

Bockemühl, J. 1964. Der Pflanzentypus als Bewegungsgestalt. Gesichtspunkte zum Studium der Blattmetamorphosen. *Elemente der Naturwissenschaft*, *1*: 3–11; reprinted in Schad, W. (ed.). 1982. *Goetheanische Naturwissenschaft. Botanik* (vol. 2, pp. 7–16), Stuttgart: Verlag Freies Geistesleben.

Bockemühl, J. 1966. Bildebewegungen im Laubblattbereich höherer Pflanzen. *Elemente der Naturwissenschaft*, *4*: 7–23; reprinted in Schad, W. (ed.). 1982. *Goetheanische Naturwissenschaft. Botanik* (vol. 2, pp. 17–35), Stuttgart: Verlag Freies Geistesleben.

Bockemühl, J. 1970. Staubblatt und Fruchtblatt. Beiträge zum Verständnis der Bildebewegung im Blütenbereich, *Elemente der Naturwissenschaft*, *13*: 12–24; reprinted in Schad, W. (ed.). 1982. *Goetheanische Naturwissenschaft. Botanik* (vol. 2, pp. 115–129), Stuttgart: Verlag Freies Geistesleben.

Brady, R. H. 1987. Form and cause in Goethe's morphology, in Amrine, F., Zucker, F. J., and Wheeler, H. (eds.). *Goethe and the Sciences: A Re-Appraisal* (pp. 257–300), Boston, MA: Springer.

Bowman, B. 2011. Goethean morphology, Hegelian science: Affinities and transformations. *Goethe Yearbook*, *XVIII*: 159–181.

Coen, E. S. 2001. Goethe and the ABC model of flower development. *Life Sciences*, *324*: 523–530.

Coen, E. S. and Meyerowitz, E. M. 1991. The war of whorls: Genetic interactions controlling flower development. *Nature*, *353*(6339): 31–37.

Dornelas, M. C. and Dornelas, O. 2005. From leaf to flower: Revisiting Goethe's concepts on the "metamorphosis" of plants. *Brazilian Journal of Plant Physiology*, *17*(4): 335–343.

Förster, E. 2001. Goethe on "Das Auge des Geistes." *Deutsche Vierteljahresschrift für Literaturwissenschaft und Geistesgeschichte*, *75*: 87–101.

Förster, E. 2011. Wie wird man Spinozist?, Die Methodologie des intuitiven Verstandes, in *Die 25 Jahre der Philosophie* (pp. 87–119 and 253–276), Frankfurt: Klostermann; trans. (2012) How to become a Spinozist?, The methodology of the intuitive understanding, in *The 25 Years of Philosophy* (pp. 75–99, 250–276), Cambridge MA, and London: Harvard University Press.

Goethe, J. W. 1790. Die Metamorphose der Pflanzen, in *Goethes Werke. Hamburger Ausgabe. Naturwissenschaftliche Schriften I* (vol. 13, 64–101), Munich: C. H. Beck 1999.

Goethe, J. W. 1798. Die Metamorphose der Pflanzen (Gedicht), in *Goethes Werke. Hamburger Ausgabe. Gedichte und Epen I* (vol. 1, 199–201), Munich: C. H. Beck 1999.

Goethe, J. W. 1816/17. *Italian Journey*, New York: Schocken 1969.

Goethe, J. W. 1817a. Glückliches Ereignis, in *Goethes Werke. Hamburger Ausgabe. Autobiographische Schriften* (vol. 10, 538–542), Munich: C. H. Beck 1999.

Goethe, J. W. 1817b. Der Verfasser teilt die Geschichte seiner botanischen Studien mit, in *Goethes Werke. Hamburger Ausgabe. Naturwissenschaftliche Schriften I* (vol. 13, 148–168), Munich: C. H. Beck 1999.

Goethe, J. W. 1952. *Goethe's Botanical Writings*, trans. B. Müller, introd. Ch. J. Engard, Honolulu, HI: University of Hawaii Press.

Heusser, P. 2008. Goethes Verständnis von Naturwissenschaft. *Goethe Jahrbuch*, *125*: 110–121.

Hindrichs, G. 2011. Goethe's notion of an intuitive power of judgment. *Goethe Yearbook*, *XVIII*: 52–65.

Nassar, D. 2011. "Idealism is nothing but genuine empiricism." Novalis, Goethe, and the ideal of romantic science. *Goethe Yearbook*, *XVIII*: 67–95.

Reill, P. H. 1986. *Bildung*, *Urtyp* and polarity: Goethe and eighteenth-century physiology. *Goethe Yearbook*, *III*: 139–148.

Seamon, D. and Zajonc, A. (eds.). 1998. *Goethe's Way of Science. A Phenomenology of Nature*. New York: State University of New York Press.

Steigerwald, J. 2002. Goethe's morphology: Urphänomene and aesthetic appraisal. *Journal of the History of Biology*, *35*: 291–328.

Steiner, R. 1897/[5]1921. *Goethes Weltanschauung*. Berlin: Philosophisch-Anthroposophischer Verlag.

Wellmon, Ch. 2010. Goethe's morphology of knowledge, or the overgrowth of nomenclature. *Goethe Yearbook*, *XVII*: 153–177.

9 'Biologie'

Lamarck's endeavor of a science of living entities[1]

Snait Gissis

Introduction

I believe that it is helpful to look at the ongoing endeavor of Lamarck from around the turn of the eighteenth century until the beginning of the 1820s as if it were a complicated piece of music composed around a core of a few bars whose contents silently herald: 'Biology exists. It has an object. It has a method. It is a science.'

Let me start with a broad generalization. Four interconnected but distinct scientific objects emerged at the end of the eighteenth and the beginning of the nineteenth centuries: 'society/the social,' 'living nature,' 'the self,' and 'race.'[2] Throughout the nineteenth century these four scientific objects were analyzed, articulated, elaborated upon, diversified and deployed as a focus of empirical research and theoretical investigations, and gradually crystallized into bounded scientific fields, into disciplines, became academized and institutionalized. It was also the case that these four objects became intertwined in myriad ways throughout that period.

Nevertheless, I would argue:

1 that, in the earlier period within the time frame of the present chapter, there were commonalities of discontent among those investigating the phenomena related to 'living entities' and those looking into 'social' phenomena[3] rather than those pertaining to state-political frames and to households.

At the roots of this common disaffection were the limited powers/abilities of contemporaneous explanatory models to deal with time-dependent processes and the historical dimensions of 'nature,' with the dynamics of whole–part relations, with modes of cohesiveness, and with the notion of functions changing with time.

2 that it should be noted that investigating 'living nature' was considered a combined empirical/theoretical endeavor right from its inception as a separate field of investigation. Later on in the nineteenth century this gave the field a privileged position vis-à-vis the three others and helped shape a context, both scientific and social-political-cultural, that allowed the assumption that there was a fundamental correspondence between organic nature and social life, as well as mental life, and thus between their

> mechanisms of development. Later, this correspondence became embedded in evolutionary features, including modes of heredity, and in the foundational entities and the types of lawfulness in these domains. However, in the time period that I deal with the interaction between various discourses was 'multidirectional traffic.'

It is within this general framework that I suggest looking at Lamarck's *biologie* as a project that gradually evolved, became actualized in all his writings from the turn of the century on, and crystallized in his concepts and mechanisms related to his transformism, their consequences and implications.

During the past 40 years of making use of a different understanding of contextualization, a new historiography, in particular one not of precursors and anticipators,[4] has been established that has yielded a continuous narrative on the emergence of 'evolution,' and thus includes Lamarck's endeavor in its late eighteenth- and early nineteenth-century contexts. My effort to sketch these 'silent bars,' mentioned in the opening sentence, is to be viewed as relating in particular to those historians and philosophers of biology who worked specifically on *biologie*, not just as a particular short essay assumed to be written around 1800–1802[5] – whose mission and projected plan were denied, retracted and partially reinstated in the years that followed – but as a continuous persevering feature of Lamarck's work from 1800 on.[6]

I have divided this chapter into two sections. In the first I shall reiterate very succinctly my argument that, when interpreting Lamarck's evolving endeavor, and in particular his transformism, by reconstructing the natural history contexts of 1780–1810 and analyzing his complicated and changing positions within them and vis-à-vis contemporaneous controversies, one should also look at the interactions and transfers from 'the social.' In the second section I briefly look into the principal interpretations that have been proffered in an attempt to crystallize commonalities and delineate disagreements on the object, the method and the nature of that study of living entities which was at times called the science of '*biologie*.' The issues that Lamarck worried about and tried to address in his works on living entities from the turn of the eighteenth century until his last work, bounded and filled his field of *biologie*. They included the chemical and physical understanding of life processes (recall that his own chemistry and physics were opposed to the new sciences of Laplace and Lavoisier), the role of hydraulic mechanisms of fluids and their 'containers,' the role of the environment, the notion of 'organization,' and his taxonomic enterprise, i.e., all embodying the flickering appearance of *biologie*.

Transfer from the social: why it is relevant for a discussion of Lamarck's conception of 'biologie'

The extensive interactions, and consequent transfers between social thought and biological thought occurred within specific social, political, cultural and institutional frameworks, and served as enabling conditions for the rise of novel conceptual frameworks. This applies in part to Lamarck's views on change within the organic world and, in particular, to Lamarck's conception of 'la marche de la nature,' i.e., his transformism. The period from the onset of the Revolution till the very early years of the nineteenth century was characterized by intense interactions among various sciences and forms of knowledge, even though the sites of these interactions had radically changed from those of before the Revolution. There have been claims to the contrary, pointing to that same period as one of growing professionalization, of increasing power of institutions and of the emergence of clear disciplinary boundaries.[7] However, it may be helpful to distinguish developments in the first revolutionary decade and those in later periods, as well as recognize parallel developments. The restructuring of frameworks created a certain number of state-funded scientific institutions and a host of state-funded educational ones in which scientists lectured/taught. These, and the double role – political and scientific – that numerous scientists played, in turn created new hierarchies, networks, clusters of patronage etc. Two such major institutions, which were founded practically one after the other, were the Museum National d'Histoire Naturelle – MHN (1793), where Lamarck worked and taught, and the Institut National des Sciences et des Arts – Institut (1795), to whose first class – 'for mathematical and physical sciences' – Lamarck was soon nominated. Primarily, until 1795, Lamarck was actively participating in a wide range of activities, all of which were connected to innovations of the new revolutionary regime or to results of the nationalization of church and private property dealing with 'natural history.'

Given the instability, not to say turbulence, inherent in that period there was a constant need to be attentive to rapidly shifting governmental practices, ideologies, and rhetoric concerning scientific practices and theorizing, as well as to the shifting of power within institutions, and in particular those of natural history. Furthermore, as Napoleon's power became all-pervasive and the influence of the reinstated Catholic Church again became extensive, the change in the regime at large and its cultural preferences had its effect on Lamarck. After publishing his theoretical work of 1800–1802 on living nature, which included his *Biologie* (1800), his *Système des animaux sans vertèbres* (1801), his *Hydrogéologie* (1802) and his *Recherches sur l'organisation des corps vivans* (1802), Lamarck's developing position could be found elaborated in the introductions to the courses he taught at the MHN, in hints in his articles on fossils that were published in the Muséum's *Annales* between 1802 and 1806, and then in the *Philosophie zoologique* of 1809. Later relevant works for this discussion include the items for the *Nouveau dictionnaire d'histoire naturelle* (1817–1818), edited by Virey,[8] his *Histoire naturelle des animaux sans vertèbres* (1815–1822) and his *Système analytique* (1820).[9] Two clusters of concepts that were prevalent in the contemporaneous sociopolitical discourses are relevant and manifest themselves in Lamarck's works:

1 the notion of processes as part and parcel of the explanatory model; and
2 the view that diversity and complexity are inevitable components of the object to be explained.

These also meant that causal explanations had to take into account filiations of events in historical time, and that a greater emphasis was placed on comparison and observation. In Lamarck's work of the late 1790s, i.e., before the transformist turn, a distinction between living and nonliving entities appeared in which the concept of organization played a subtle role, at first in distinguishing between living and nonliving, and somewhat later in drawing the line between different sorts of living entities. This coincided in interesting ways with the 'organizational' discourse and rhetoric of both the revolutionary period and the early Thermidorean, post-Jacobin regime. The malleability of individuals and of societies was assumed to be bounded by 'structural' constraints whereby individual and social behavior were conceived as revealing organizational regularities, even as emerging ones. Thus, though 'perfectibility' and 'progress' were presented as open-ended, they were bounded by the assumptions on possible horizons. Let me add that, though the Jacobins acted as if they had created 'a break in time,' they pictured whatever came afterwards as an ascending, gradual process of progress (Gissis 2009). In almost every introductory lecture from 1803 on, Lamarck, whose starting point was 'the panorama of living nature' rather than 'Man,'[10] called upon his listeners to exercise their imagination in order to encompass the vast variety and diversity of living nature. In 1806 he actually suggested to his audience the following thought experiment (repeated almost verbatim in later texts): Suppose that the whole of nature were laid out for a spectator to contemplate as if it were a collection; could one then classify nature in a way that 'would cut it really at its joints,' reveal 'les vrais rapports' that 'really' describe 'la marche de la nature' (Ancien Cours 1806, Inédits). In numerous introductory lectures Lamarck started from a classificatory scheme, asking what could count as a common feature, thus calling upon his students to delve more deeply into the foundations upon which one could reconstruct the natural *rapports*. Regarding foundations, individuals formed the axis of the contemporaneous discourse on the social. In fact, the movement to viewing society either as some kind of aggregate or as an emergent collectivity was related to choosing the social model of progress, which in almost all its then-current variations in France was based on the ontological primacy of the individual. When Lamarck's transformist notions became much more clearly formulated, the concept of the individual and its role within his system crystallized as well. The lowest boundary of the living entities was also the almost inconceivable limit of animalization, and thereby of living structure (introduced through spontaneous generation). However, it would be wrong to say that there was 'more or less individuation.' It was in relation to this end point of the continuum of the living – of the animalized, the individualized – that the role of time became manifest. Time was that through which the gradual emergences and changes occurred, and through which the ordering took place in an environmentally causal manner, and this via particular mechanisms of change, specifically those of use and habit. A proper individual was a living entity which bore resemblances to other individuals, and this collection of individuals constituted a species. This was sharply formulated in his *Biologie* of 1800: 'Un corps vivant est un corps naturel borné dans sa durée, dans ses parties au moins dans les principales, possédant ce qu'on nomme la vie.... *Ce corps, constitue un individu*...' (Lamarck 1800, 7, my emphasis). Given that the entities that changed were primarily individuals, and qua individuals given their progressively complexified structures, and given that individuals could be considered as belonging to a species by virtue of secondary consequences, one can see how with transformism the attribution of reality was gradually transferred from the species to individuals in the period from 1800 to 1803.[11] However, given that both the environment and the living entities within it were in a constant process of change and in constant interaction, stability resided primarily in the pattern of the process rather than in its contents.

Lamarck's innovations received various explanations and interpretations, mostly located in the natural history and physical sciences communities of the time, and can be found in their problématiques, debates and controversies.[12] One should add to this picture the social discourses and practices of that time. I believe that the problématique that Lamarck dealt with was certainly that of the natural history communities of the pre-revolutionary past and that of the contemporaneous ones, but his innovative solutions should *also* be understood as resulting from the transfers of notions, concepts, models and metaphors from the social discourses. He transferred a model of progress, variations of which could be found not only in the actual writing of prominent contemporary figures but also under various guises and in odd combinations in the public sphere of culture and laws and the administrative discourse. By combining time, genealogy and causality, such a model could provide an analysis of processes that could describe the emergence of complexity that Lamarck came to see as the principal feature of living entities. It was not a straightforward and simplistic transposition but rather a deployment of the elements, transferred and indirectly translated, within his idiosyncratic views of natural history, changing their referents and their role in the overall narrative constructed thereby.[13]

In conclusion, I think that the possibility of transfer from contemporaneous social discourses impacted on Lamarck's crystallizing notions on/of the object of his endeavor – living individual entities and their prominent features – thereby helping in the setting the boundaries of the field of *biologie*, laying down certain fundamental features of it and sketching what makes it possible to 'construct' living nature.

Lamarck's endeavor of *biologie*: the Spectrum of Commonalities and disagreements among contemporary interpreters

Making a wide generalization, I would say that, by the time Lamarck was constructing his field of *biologie*, doing science had already meant both gathering the plurality of phenomena under laws of universal/regularizing application and decomposing, from the whole to the parts, seeking solutions lower and lower in the hierarchy of levels, quantifying, reaching unto ever smaller entities.[14] In terms of 'biological temperament,' this could be expressed in an ontology of dynamic processes or in an ontology of stable, nonchanging entities. Lamarck deployed the tool of decomposition over his living entities but refused to quantify and to measure[15] the stages of these processes in the manner advanced by the 'new sciences' and their 'new language' epitomized for him in the figures of Lavoisier and Laplace (Burlingame 1973; Burkhardt 1977; Corsi 1988; Riskin 2002). He chose dynamic processes over 'dumb,' static entities. He also does not fit neatly into the more conventional narrative in biology of the tension between natural history and laboratory physiology, which posited the latter as the one seeking to emulate physical sciences and seek universalistic knowledge. One could argue that this dichotomous division fits better the somewhat later nineteenth-century positions and practices rather than the checkered landscape of the late eighteenth and very early nineteenth centuries, all the more so in recalling the positions then taken by the young Geoffroy, Lamarck and Cuvier concerning field, cabinet and comparative laboratory practice, respectively. Lamarck did not experiment, though there are indications that he followed and was acquainted with at least some of the experimental work of his immediate predecessors and that of his contemporaries. From early on, even before the crucial turn to transformism, Lamarck had looked upon living entities as distinct and this not by virtue of the makeup of their molecules, and thus without infringing (his) physical-chemical assumptions of the makeup of matter. His later constructions of the makeup, the causalities involved, the evolvement and history of the plurality and the diversity of living entities did not change that position.

Almost all contemporary writings on Lamarck's endeavor regarding *biologie* agree that understanding Lamarck's deployment of his physics and his chemistry is crucial. They also agree that his 'biology' views were closely intertwined with his views on the *geological history* and *structure* of the 'crust of earth,' thus employing the differential time scales involved with each of the scientific fields (see Gohau 1997, 2006; Laurent 1987; Rudwick 2005, 2009). Some commentators emphasize the rejection of the Cartesian mechanistic model on the one hand and on the other the adoption of (the specifically late eighteenth-century) Newtonian model, in a manner described as analogous, or analogously (Conry 1981; Corsi 1982; Roger, particularly 1979; Wolfe 2014).

Assumed, implied or explicitly stated is the following: In order to understand Lamarck's weighty apparatus, his innovative stance, as well as the unresolved inner tensions within his theoretical framework, one has to relate his expository efforts to the overall gradually evolving conceptual scheme he had devised, as well as to the *changing contemporaneous natural history problématique*, such as questions concerning the conditions for the existence of living entities; the attributes of the material that living entities were constituted of; whether there were common functions, common features of all living entities; and the properties of the environments of living entities that made their existence and reproduction possible. This, it is argued, will help clarify the presuppositions and theoretical constraints with which he struggled. One also has to relate Lamarck's framework to the prevalent contemporaneous conceptualizations of physics and chemistry. It is argued that the Newtonian tradition in 'natural philosophy,' i.e., in the physical sciences, served in a double role, namely as a basic conceptual scheme and as a paradigmatic model of how to do science and what a scientific field should look like (Biener and Schliesser eds., 2014). Also that one of the features of the Newtonian tradition is relevant, namely that the scientific search was for laws of nature that had universal extension, had universal applicability and were to account most generally for the totality of phenomena as regulated under them. These dynamical laws were assumed to hold at all levels. However the 'individuals/entities' dealt with in Lamarck's *biologie*, contrary to those in the Newtonian model, had history, carried history within their varying organization, were transforming, were nature-produced (and not God-produced), and their interactions with their specific environments were neither determined nor unchanging.

There is also a common agreement regarding the centrality of understanding Lamarck's use of fluids, – subtle fluids, imponderable fluids – which within his conception (and, more generally, within those investigating and making use of them during the second half of the eighteenth century), whether 'heat,' 'caloric' or 'electricity,' assumed that they could expand, transfer physical material attributes, and thus act as causes/be a locus of causality, but could not be translated quantitatively into properties of microscopic, foundational entities. In the general eighteenth-century debate on the assignment of bodily structures and mechanisms of living entities there was a broad choice of models. Lamarck, in line with his geological and physical views, chose for his composition of living entities[16] a hydraulic model for the dynamics of various active fluids as transferring agents. These fluids also molded and gave form to the body structures and traveled in pipes, tubes, canals and other equivalent spaces. Organization was thus seen as the model's necessary consequence. These fluids functioned as connections between various parts of the inside as well as linkages between the inside and the outside. More generally, they enabled the 'rapports' between the environment and the body. All interpreters agree that it is within specific parts of these structures of living entities that the transformation into signals of inner and outer cues was being carried out.[17] However, as is sharply brought out by Corsi (1988, Chapter 5, 2006), within a few years Lamarck's emphasis had moved from elaborating the generalized dynamics of atmospheric and terrestrial fluids and by analogy those of living entities to more specific zoological issues.

Lamarck's chemistry, in particular, is seen as playing a predominant role in his conceptualization of both geology and meteorology. Lamarck's conceptions on the nature of 'the crust of the earth' and on '(external) environment' (including what he formerly called 'climate') are in turn looked upon as crucial to understanding his view on the possibility of delineating a field of living entities – *biologie*. There had been an almost unanimous agreement in the scholarship on Lamarck's rejection of the 'new sciences' and in particular the concept of chemical analysis, an agreement shared by those who offered close readings, analyses and interpretations of both his early and his later works, this starting with Schiller[18] and Hodge to the latest articles by Corsi. However, recently a different interpretation has been proffered by Angela Bandinelli (2013). Bandinelli offers an analysis of Lavoisier, on the basis of which she reconstructs Lamarck's conceptualizations as moving gradually from that position combined with a conception of an essential molecule, toward his transformism, which also entailed his changing his view of the relations between the inorganic and organic. Up to this point it is very likely that other interpreters would largely concur. However, the crucial change in terms of Lamarck's attitude to the new chemistry occurred according to Bandinelli mostly between 1802 and 1809, and was made evident in the *Philosophie zoologique* and further detailed in the first volume (1815) of his *Histoire naturelle*. She argues that Lavoisier offered a novel perspective on living entities, a unifying-uniforming conception of both the organic and the inorganic, which led to a view of living entities as combustible bodies[19] and resulted in the discarding of the view of living systems as machines. Lamarck in his earlier phase had differentiated between the principle of 'power of life' and 'matter.'[20] In her interpretation Lamarck's view of *biologie* evolved into a physical-chemical, even physiological, science that was not dependent on mechanics and accepted and even followed Lavoisier's view that living entities (as well as nonliving ones) were 'natural chemical systems' in a process of constant material transformation with their external environment. I believe that what is implied by this interpretation is that it was through the learning of this new viewpoint that Lamarck also adopted the idea that compounding was capable of producing novelty.

The spectrum of positions on the meaning and significance of both 'milieux' and 'milieux environnans' is not very wide since there is general agreement on the crucial role of these 'environments' in impacting the course of action of the powers that are involved in shaping and molding living entities within a transformist framework. Also agreed upon is that this notion of 'milieux' is closely interwoven with that of 'change' and of differential 'interaction' whereby elementary living entities are passively and directly exposed while the increasing complexity of living entities can be regarded as an index of the degree of their mediated constant interactions with their milieux, through 'needs' and 'habits.' Thus, milieux, together with the 'power of life,' are perceived as catering for, constructing and nurturing variations. Most historians agree upon relating the branching of the rising orders of living entities to the impact of these various interactions, with their dynamics represented by the action of fluids inside and outside the living entities.

A note on 'organization' is called for here. There is general agreement on the centrality of the concept and its accompanying concept, 'function.' There is, however, disagreement on its deployment. The controversy has to do with respective positions on 'the physical' and 'the chemical' in Lamarck's oeuvre and their relationship with the notion of organization. The positions expressed cover a wide range. At one end, viewing 'organization' and self-organization as rich notions that allowed for a holistic mode of explanation alternative to the then prevalent mechanistic one, notions through which living entities were defined (regarding self-organization see e.g. Conry 1980, 1994; Jordanova

1989). At the other end there is an array of interpretations which explicate 'organization' primarily as a central concept in Lamarck's theory of transformism, a concept that turned the naturalist's gaze from external characteristics of living entities to their internal organization (e.g. Burkhardt 1977). Alternatively, it is looked upon as deployed to draw hierarchical-organizational distinctions among living entities, and concurrently was used to unify the whole field of living entities within a single theoretical frame. There 'the organic' is conceived as organizational, whose foundational causality was physical, as exemplified by the motion of fluids.[21]

All interpreters note and comment on the discrepancy between Lamarck's contemporaneous renown as a classifier and innovative taxonomer and what seemed for a long time in the historiography to be his marginalization as an innovative naturalist. However, in the past three decades a great deal of innovative work has appeared that has reinstated Lamarck as a member of multiple communities, inside as well as outside the more researched institutions such as the MHN and the Institut. The spreading of his ideas to many corners of Europe, their complex assimilation and deployment also by virtue of the courses he taught at the MHN has been reconstructed, and the interlocutors of his *dialogues manqués* have been resurrected.[22] Furthermore, recent historiographic work on Cuvier has complexified and brought into focus the controversies about the proper sites and the proper methods for practicing natural history, considering not only the personal elements but also situating Lamarck within the political and cultural contexts of the Napoleonic and Restoration periods (see Appel 1987; Laurent 1987; Outram 1984; Rudwick 2005, 2008).

The role of taxonomic work in Lamarck's *biologie* endeavor is discussed by most of this group of historians. They have pointed out that there are two ways to understand the term 'classification,' one as comprehensive and general and the other as more limited:

1. In its more limited version as meaning sorting and dividing, i.e., classifying for practical purposes, such as helping work done by naturalists and amateurs in the field in botany, a classification that was looked upon as 'artificial,' rather than rendering 'the natural order' within Lamarck's own discussions of method.[23]
2. In its comprehensive mode as referring to a *taxonomic enterprise*, which concentrated on how to move from the 'points of animaculation' constantly being produced through spontaneous generation, to living entities – first the simplest and lowest in the order of evolution and moving upward to those appearing later that were more complex. In this version, *biologie* posited the contour of an 'ideal,' as well as of 'reproductive-factual,' continuous and in principle gradual, movement of complexification of living entities, proffering both linear and branching taxonomic pictures, with both considered as rendering 'the order of "nature."'[24] Moreover, as Corsi has emphasized, this represented a mode of constituting diversity within the plurality of living entities.

The issues at the focus of contemporary disagreements have to do with the question of whether one can actually view these two activities of 'classification' and *'biologie'* of Lamarck as things apart or rather as intimately intertwined, so that arguments on taxonomy were either posited as if they substantiated the theoretical claims at one stage of his work, or else exemplified such claims at a later stage. Answering this question is complicated by the fact that Lamarck did not experiment, or reconstruct experiments, or dwell at great length on accumulations of empirical data comparisons.

Let me add the following comment:[25] in one of the rare semiautobiographical paragraphs in an opening course lecture Lamarck stated (1816) that only gradually did he realize what a rare glimpse into the working of nature the investigation of invertebrates had offered him.[26] He looked upon the distinction between invertebrates and vertebrates as one of his important original contributions. It effectively changed both the boundaries of and the perspective on the realm of animals he had been nominated to investigate at the MHN, and if we follow Lamarck in his works we find that in some sense, he tended to posit the invertebrates as 'model organisms.' In 1806, in the introductory lecture to his course on invertebrates, Lamarck suggested a new definition of living entities based on the most general feature found in the lowest forms of organization, rather than on the most sophisticated ones. The evolutionary perspective thus became the foundational one through which one classified. The predominant function that served as a tool of that classification was the presence or absence of nervous activity, signaled by the presence or absence of any form of a specialized system of nerves. One could argue that Lamarck endeavored to turn the *Histoire Naturelle des Animaux sans Vertèbres* into an exposition of 'evolutionary systematics' – with a heavy emphasis on 'evolutionary' – because he probably realized that the book would be of interest to a wide audience. The principal thrust of the general part was presented in the form of 'first principles' in which Lamarck tried to tie together method, model and a rhetoric adequate to the task, the emphasis being on the fact that this framework worked adequately, fittingly, all along the evolutionary path[27] between organisms and their subsystems and faculties and their 'needs.'[28]

Lastly, most of the recent authors trace the appearance of Lamarck's project on *biologie* and date it to either 1800 or else to early 1802.[29] There is general agreement on the chronology of the appearance of the term and the project related to it, i.e. *Biologie* (1800), 'biologie' declared as 'terrestrial physics' in the final part of *Hydrogéologie* (1802a). There is also a consensus on Lamarck's later decision to discontinue it, claiming reasons such as age, failing health, preoccupation with other projects in the *Recherches sur l'organisation des corps vivans* (1802b), the declared repetition of that decision in the *Philosophie Zoologique* (1809), and his very partial, somewhat declarative, return to the project, as indicated by his usage of *biologie* and its derivatives, such as 'biologiste,' in the first volume of *Histoire Naturelle des animaux sans vertèbres* (1815) and its quick withdrawal from further volumes of that oeuvre. A limited number of explanations have been offered why this was so, these being either mostly internalist or in terms of his relations with the naturalist community. However, Corsi's contextualization of the *biologie* project and in particular of Lamarck's rhetorical strategies and methodological choices throughout the 1800–1822 period has added another dimension to our understanding as he has interlaced Lamarck's actual use of the term and the description of his project with the fast-changing political regimes and the fortunes of 'materialist' positions, of 'God,' of 'the Catholic Church,' in the politics and culture of Napoleonic France and those of the early Restoration period (Corsi 2012).[30]

Concluding remarks

Lamarck lived in a period of transition, of shifting boundaries and fast-changing conceptual schemes. Two of the principal scientific institutions to which he belonged were initially conceived as sites for the interaction and mingling of scientific discourses, and continued to be such sites to some extent while undergoing reorganization. To understand Lamarck the picture needs to be refined, distinguishing developments in the first revolutionary decade and those in later periods, and distinguishing between parallel developments – such as the mingling of discourses and the professionalization of the sciences – which took place side by side for a while, and by analyzing the political–economic–cultural context that made these possible. Politically, the first decade of the Revolution, – i.e., until Napoleon declared himself first consul (in the year 1801) and three years later emperor – was marked by political instability – of one kind until 1795, and of another later. The restored Crown in 1814 also changed hands within the next ten years. Within this context the methods and 'the language of sciences' changed rather dramatically. In some ways Lamarck was a product of an earlier period, yet he offered an innovative conceptual scheme, and deftly mixed different time scales within that scheme. It began to be more clearly formulated in his writings around the turn of the eighteenth century, while using concepts, mechanisms and methods that he laid down earlier, which were part and parcel of earlier modes of thinking. When he formulated this innovative theory on a grand scale, encompassing the whole of living nature, the scientific field in particular and the cultural field at large were beginning to undergo a transformation that eventually hardened the disciplinary boundaries.

His audiences were to be found both in scientific and nonscientific communities, mostly outside the two institutions to which he belonged during that period – the Institut and the MHN. This, combined with his idiosyncratic mode of writing, opened up many modes of reading and interpreting him already during his lifetime. These were enabling conditions for later dissemination of severed components of his enterprise, and their incorporation into theories at times rather incompatible with his own, but it also ensured the many long lasting transformations of some of his principal conceptions.

His innovative endeavor of constituting a science of transforming living entities – *biologie* – received various explanations and interpretations, which mostly centered on the natural history and physical science communities. In the first part of this chapter I argued that it would be relevant to add to this picture the contemporaneous social discourses and practices of that time. In the second part I briefly indicated the principal foci of historical-conceptual agreements and controversies centered on various facets of his endeavor and its possible sources, as well as around the possibility of a view of his taxonomic and his evolutionary enterprises as an integrated framework – his *biologie*.

Notes

1 The chapter is dedicated to the memory of (Silvan) Sam Schweber (1928–2017), rara avis.

2 See e.g., Daston 2000, 1–14, and Baczko 1994; Gissis 2010, 2011; Goldstein 2000, 2005; Heilbron 1995; Jones 2002; Rey 1994; Wagner 2000; Wahrman 2004; Wheeler 2000.

3 On the gradual conceptual separation of the 'social' first as an adjective (in Rousseau's work) and later on as a noun – society – denoting something which is separate from the state, even though the state's boundaries were its boundaries, see e.g., Baker 1973, 1990, Carrithers 1995, Heilbron 1995, Staum 1996.

4 Contemporary issues in the field of life sciences have contributed to a renewed interest in Lamarck's work, e.g., nongenetic modes of inheritance, the role of development, the importance of environment, and twentieth-century notions of self-organization and of complexity, e.g., Gissis and Jablonka eds. 2011.

5 As discussed in the first section.

6 In particular in alphabetic order: Badinelli, 2013; Barsanti, 1995, 1997a, 1997b, 2000; Barthélemy-Madaule, 1982; Bourdier and Orliac, 1971; Burkhardt, 1977, 1987, 1995; Burlingame, 1973, 1981; Conry, 1980, 1981, 1994; Corsi, 1982, 1988, 1997, 2005, 2006, 2012; Dupuis, 1997; Gillispie, 1997; Gohau, 1997, 2006; Gould, 1999a, 1999b; Goux, 1997; Hodge, 1971, 1995; Jordanova 1984, 1989; Laurent, 1987; Mayr, 1972; Omodeo, 1971; Roger, 1979, 1995; Russo, 1981; Schiller, 1969, 1971; Szyfman, 1982.

7 Such change can be traced through the changes in the structure of the Institut after the turn of the century, also through the structure of Cuvier's report of 1810. See, e.g., Crosland 1992, Dhombres and Dhombres 1989.

8 Note that all the nonspecialist items written by Lamarck (i.e., not the ones on conchology and on meteorology) were followed by parallel entries by Virey offering converse/obverse positions. See also Corsi 1987.

9 See items in P. Corsi, general editor: www.lamarck.cnrs.fr/index.php?lang=fr, as well as Lamarck 1812, 1907, 1933, 1972, 1990.

10 Recall that for Lamarck the faculty of reason was looked upon as causally dependent on matter, and matter being the sole entity the naturalist could investigate and explicate. However, for a strong argument for the primacy of death in Lamarck's material universe, see Giglioni in Normandin and Wolfe eds. 2013. See also Wolfe 2016, esp. Chapters 5–6.

11 Whether considered from the perspective of the order of becoming or that of knowing, in both the individual became the unit of the analysis, with 'masses' the principal category of ordering. 'Masses' were nonclassified collections of individuals that fell under very generalized divisions of the living entities, viewed from the point of view of combined timelines, of diversified environmental conditions and complexifying organizational form lines (e.g., already in the *Discours Préliminaire*, 1802).

12 I use the term 'problématique' to delineate a space of clusters of questions rather than lists of topics, questions posited, debated, solved and resolved, considered as legitimate or illegitimate, where the boundaries of such a space are changing and

context-dependent, while its core questions may be more stable. Such clusters can be and are answered in differing modes, e.g., metaphysically, methodologically, experimentally.

13 These issues are discussed in great detail in Gissis 2009, 251–270.

14 Laplace, whose scientific work and institutional politics Lamarck had shunned, was considered the outstanding contemporaneous representative of such a 'scientist.' See e.g., Fox 1990, 278–294, Gillispie *et al.* 1997.

15 When looking at his use of both 'mesurer' and 'quantité,' it seems that the latter was rare indeed. When 'quantité' was used, it referred mostly to an undistinguished large number of the units and/or the matter discussed, e.g., acts of nature, rain, electricity; 'Mesurer' was rarer still, mostly in earlier works relating to meteorology, or later on to irritability.

16 The presence or absence of a bodily system which enabled the activity of the various body fluids, such as blood, was considered a distinguishing feature, marking the whole order of invertebrates, and used later on for internal classification within that order. See also discussion in Cheung, Chapter 6, this volume.

17 Already in the appendix to the *Recherches* (1802b), and in line with the functioning of passivity, the basic tenet of sensationalism, for Lamarck the active agent in sensations, in feelings and in volition is not the nervous system nor any part of it – they are the loci of the events and processes! – but a particular fluid: 'le fluide nerveux.' Forming habits meant changing the flow of these fluids. See also Gissis 2010.

18 Though note Schiller's emphasis on the emergence of 'a new science' in his 1978 book.

19 Clearly challenging Holmes's analysis on these issues, see especially Holmes on Lavoisier and the Chemistry of Life (Holmes 1985).

20 Note though that Barsanti, who argues for a holistic view of Lamarck's endeavor of 'biologie' already has a few statements indicating that direction (2000, 126–127).

21 See e.g., Roger; the intricacies of relations between organism and organization are brought out in Cheung 2006.

22 In particular Corsi, but also some of the others.

23 See the long discussion by Burlingame, as well as Schiller 1969, 1971, both relating to Daudin's work, 1983 [1926]. The division into 'natural' and 'artificial' is discussed e.g., in Farber 1981; Sloan 1995.

24 The issue of 'branching' was dealt in a particular manner by Gould 1999a, 1999b.

25 See also Gissis 2010.

26 'il existoit réellement un ordre à reconnoitre, relativement à la composition de chaque sorte d'organisation animale' (les Inédits, 29).

27 *Histoire*, Intro. 23–24. This had an underlining pointing at Cuvier's conception of 'nature,' intending to undermine the latter's definition for 'animal.' Lamarck's own exposition of the characteristics of a 'living body' harked back to the project 'Biologie.'

28 Yet, in the supplement to that work, Lamarck admitted that the actual order, or 'the order of productions,' as he named it, was differently constructed.

29 Already Hodge 1971; contrary to its much later dating by the first editor and publisher – Grassé 1944.

30 Corsi also points out that Lamarck used the same pattern when he explained (away) putting an end to his meteorology project in 1810.

References

Appel, T. A. 1987. *The Cuvier-Geoffroy Debate: French Biology in the Decades before Darwin*. New York: Oxford University Press.

Baczko, B. 1994. Enlightenment and the institution of society: Notes for a conceptual history. In W. Melching and W. Wyger (eds.), *Main Trends in Cultural History: Ten Essays* (pp. 95–120). Amsterdam: Rodopi.

Baker, K. M. 1973. Politics and social science in eighteenth-century France: La société de 1789. In F. Bosher (ed.), *French Government and Society 1500–1850: Essays in Memory of Alfred Cobban* (pp. 208–230). London: Athlone Press.

Baker, K. M. 1990. *Inventing the French Revolution: Essays on French Political Culture in the Eighteenth Century*. Cambridge, New York: Cambridge University Press.

Bandinelli, A. 2013. *Le Origini Chimiche della Vita: legami tra la rivoluzione de Lavoisier e la biologia di Lamarck* (pp. 133–200). Biblioteca di Nuncius, Studi e Testi LXXI. Florence: Leo S. Olschki.

Barsanti, G. 1995. La naissance de la biologie: Observations, theories, métaphysiques en France 1740–1810. In C. Blanckaert, J. L. Fischer and R. Rey (eds.), *Nature, Histoire, Société: essais en homage à J. Roger* (pp. 197–227). Paris: Klinchsieck.

Barsanti, G. 1997a. Lamarck et la naissance de la Biologie. In G. Laurent (ed.), *Jean Baptiste Lamarck 1744–1829* (pp. 349–368). Paris: Editions du CTHS.

Barsanti G. 1997b. Le 'Système de la nature' de Lamarck (1794): analyse d'un ambitieux projet avorté d'après un manuscrit oublié. In C. Blanckaert, C. Cohen, P. Corsi and J-L. Fischer (eds.), *Le Muséum au premier siècle de son histoire* (pp. 219–228). Paris: Éditions du Muséum national d'histoire naturelle.

Barsanti, G. 2000. Lamarck: Taxonomy and theoretical biology. *Asclepio*, *LII*(2): 119–131.

Barthélemy-Madaule, M. 1982. *Lamarck, the Mythical Precursor: A Study of the Relations between Science and Ideology*. Trans. by M. H. Shank. Cambridge, MA: MIT Press.

Biener Z. and Schliesser E. (eds.). 2014. *Newton and Empiricism*. Oxford, New York: Oxford University Press.

Bourdier, F. and Orliac, M. 1971. *Esquisse d'une chronologie de la vie de Lamarck*. Paris. École Pratique des Hautes Études.

Burkhardt, R. 1977. *The Spirit of System: Lamarck and Evolutionary Biology*. Cambridge: Harvard University Press.

Burkhardt, R. 1987. Lamarck and species. In S. Atran *et al.* (eds.), *Colloque international Histoire du concept d'espece dans les sciences de la vie*. Paris: Editions Singer Polignac.

Burkhardt, R. 1995. Lamarck in 1995. In *The Spirit of System: Lamarck and Evolutionary Biology* (pp. i–xl). Cambridge, MA: Harvard University Press.

Burlingame, L. J. 1973. *Lamarck's Theory of Transformism in the Context of His Views of Nature, 1776–1809*. Dissertation. Cornell University.

Burlingame, L. J. 1981. Lamarck's chemistry: The chemical revolution rejected. In H. Woolf (ed.), *The Analytic Spirit* (pp. 37–64). Ithaca and London: Cornell University Press.

Carrithers, D. 1995. The Enlightenment science of society. In C. Fox, R. Porter and R. Wokler (eds.), *Inventing Human Science: Eighteenth Century Domains* (pp. 232–270). Berkeley, CA: University of California Press.

Cheung T. 2006. From the organism of a body to the body of an organism: Occurrence and meaning of the word 'organism' from the seventeenth to the nineteenth centuries. *The British Journal for the History of Science*, *39*(3): 319–339.

Conry, Y. 1980. L'idée d'une marche de la nature dans la biologie pre-Darwinienne au XIXe siècle. *Revue d'Histoire des Sciences*, *33*: 97–149.

Conry, Y. 1981. Une lecture Newtonienne de Lamarck, est-elle possible? In J. Schiller (ed.), *Lamarck et son temps, Lamarck et notre temps* (pp. 37–64). Paris: Vrin.

Conry, Y. 1994. Lamarck, Penseur de frontière. *Nuncius*, *9*(2): 559–592.

Corsi, P. 1982. Models and analogies for the reform of natural history. Features of the French debate 1790–1800. In W. Bernardi and A. La Vergata (eds.), *Lazzaro Spallanzani e la biologia del Settecento: teorie, esperimenti, istituzioni scientifiche: atti del convegno*, March 1981 (pp. 381–396). Florence: L. S. Olschki.

Corsi P. 1987. Julien-Joseph Virey, le premier critique de Lamarck. In S. Atran *et al.* (eds.), *Histoire du concept d'espèce dans les sciences de la vie* (pp. 181–192). Paris: Fondation Singer Polignac.

Corsi, P. 1988. *The Age of Lamarck: Evolutionary Theories in France 1790–1830*. Trans. by Jonathan Mandelbaum. Berkeley, CA: University of California Press.

Corsi, P. 1997. Les élèves de Lamarck: un projet de recherche. In G. Laurent (ed.), *Jean Baptiste Lamarck 1744–1829* (pp. 515–526). Paris: Editions du CTHS.

Corsi, P. 2005. Before Darwin: Transformist concepts in European natural history. *Journal of the History of Biology*, *38*(1): 67–83.

Corsi, P. 2006. Biologie. In P. Corsi, J. Gayon, G. Gohau and S. Tirard (eds.), *Lamarck Philosophe de la nature* (pp. 37–64). Paris: Presses Universitaires de France.

Corsi, P. 2011. Idola Tribus: Lamarck, politics and religion in the early nineteenth century. In S. Fasolo (ed.), *The Theory of Evolution and Its Impact* (pp. 23–25). Dordrecht: Springer.

Corsi, P., general editor of the Lamarck site on the web. (n.d.). www.lamarck.cnrs.fr/index.php?lang=fr.

Crosland, M. 1992. *Science under Control: The French Academy of Sciences. 1795–1914*. New York: Cambridge University Press.

Daston, L. 2000. The coming into being of scientific objects. In L. Daston (ed.), *Biographies of Scientific Objects* (pp. 1–14). Chicago, IL: University of Chicago Press.

Daudin, H. 1983. *De Linné à Lamarck: méthodes de la classification et idée de sèrie en botanique et en Zoologie (1740–1790)* (1926). Paris: Editions des archives contemporaines.

Dhombres, N., and Dhombres J. 1989. *Naissance d'un pouvoir: sciences et savants en France (1793–1824)*. Paris: Editions Payot.

Dupuis, P. 1997. Actualité de Lamarck taxinomiste. In G. Laurent (ed.), *Jean-Baptiste Lamarck* (pp. 369–381). Paris: Éditions du CTHS.

Farber, P. L. 1981. Research traditions in eighteenth-century natural history. In W. Bernardi and A. La Vergata (eds.), *Lazzaro Spallanzani e la biologia del Settecento* (pp. 397–403). Florence: L. S. Olschki.

Fox, R. 1990. Laplacian physics. In R. C. Olby *et al.* (eds.), *Companion to the History of Modern Science* (pp. 278–294). London: Routledge.

Giglioni, G. 2013. Jean-Baptiste Lamarck and the place of irritability in the history of life and death. In S. Normandin and C. T. Wolfe (eds.), *Vitalism and the Scientific Image in Post Enlightenment Life Science, 1800–2010* (pp. 19–49). Dordrecht: Springer.

Gillispie, C. C. 1997. De l'histoire naturelle à la biologie: relations entre les programmes de recherché de Cuvier, Lamarck et Géoffroy Saint-Hillaire. In C. Blanckaert, C. Cohen, P. Corsi and J-L. Fischer (eds.), *Le Muséum au premier siècle de son histoire* (pp. 229–240). Paris: Editions du Muséum d'histoire naturelle.

Gillispie C. C., with Fox, R., and Grattan-Guinness, I. 1997. *Pierre-Simon Laplace, 1749–1827: A Life in Exact Science* (pp. 166–175). Princeton: Princeton University Press.

Gissis, S. B. 2009. Interactions between social and biological thinking: The case of Lamarck. *Perspectives on Science, 17*(39): 237–306.

Gissis, S. B. 2010. Lamarck on feelings: From worms to humans. In C. T. Wolfe and O. Gal (eds.), *The Body as Object and Instrument of Knowledge* (pp. 211–239). Dordrecht: Springer.

Gissis S. B. 2011. Visualizing 'race' in the eighteenth century. *Historical Studies in the Natural Sciences, 41*(1), 41–103.

Gissis S. B. and Jablonka E. (eds.). 2011. *Transformations of Lamarckism: From Subtle Fluids to Molecular Biology*. Cambridge, MA: MIT Press.

Gohau, G. 1997. L'hydrogéologie et l'histoire de la géologie. In G. Laurent (ed.), *Jean Baptiste Lamarck 1744–1829* (pp. 137–147). Paris: Editions de CTHS.

Gohau, G. 2006. Lamarck philosophe? In P. Corsi, J. Gayon, G. Gohau and S. Tirard (eds.), *Lamarck Philosophe de la nature* (pp. 9–36). Paris: Presses Universitaires de France.

Goldstein, J. 2000. Mutations of the self in old regime and post-revolutionary France. In L. Daston (ed.), *Biographies of Scientific Objects* (pp. 86–116). Chicago, IL: University of Chicago Press.

Goldstein, J. 2005. *The Post-Revolutionary Self: Politics and Psyche in France, 1750–1850*. Cambridge, MA: Harvard University Press.

Gould, S. J. 1999a. A division of worms: Jean Baptiste Lamarck's contribution to evolutionary theory, part I. *Natural History* (February).

Gould, S. J. 1999b. Branching through a wormhole: Reclassifying the types of 'worms,' part II. *Natural History* (March).

Goux, J.-M. 1997. Lamarck et la chimie pneumatique à la fin du XVIIIe siècle. In G. Laurent (ed.), *Jean Baptiste Lamarck 1744–1829* (pp. 149–161). Paris: Editions du CTHS.

Heilbron, J. 1995. *The Rise of Social Theory*. Trans. by S. Gogol. Minneapolis, MN: University of Minnesota Press.

Hodge, M. J. S. 1971. Lamarck's science of living bodies. *British Journal of the History of Science, 10*: 323–352.

Hodge, M. J. S. 1995. Lamarck: Un grand changement de cadre conceptuel. In C. Blanckaert, J.-L. Fischer, and R. Rey (eds.), *Nature, Histoire, Société: Essais en Homage à J. Roger* (pp. 229–239). Paris: Klincksieck.

Jones, G. S. 2002. The new social history. In C. Jones and D. Wahrman (eds.), *The Age of Cultural Revolutions: Britain and France 1750–1820* (pp. 94–136). Berkeley, CA: University of California Press.

Jordanova, L. J. 1984. *Lamarck*. Oxford, New York: Oxford University Press.

Jordanova, L. J. 1989. Nature's powers: A reading of Lamarck's distinction between creation and production. In J. Moore (Ed.), *History, Humanity and Evolution* (pp. 71–98). Cambridge: Cambridge University Press.

Lamarck, J.-B. P. A. de Monet de. 1800. *Discours d'ouverture prononcé le 12 Floreal a l'an 8*. Lamarck website. www.lamarck.cnrs.fr/index.php?lang=fr.

Lamarck, J.-B. P. A. de Monet de. 1800/1944. *Biologie ou Considérations sur la nature, les facultés, les développemens et l'origine des corps vivans*, ed. P. P. Grassé. *La Revue Scientifique, 82*: 267–276.

Lamarck, J.-B. P. A. de Monet de. 1801. *Système des animaux sans vertèbres; ou, tableau général des classes, des ordres, et des genres de ces animaux précédée du discours d'ouverture du cours de zoologie, donné dans le Muséum national d'histoire naturelle l'an 8 de la République*. Paris: Deterville.

Lamarck, J.-B. P. A. de Monet de. 1802a. *Hydrogéologie, ou, Recherches sur l'influence qu'ont les eaux sur la surface du globe terrestre*. Paris: Chez l'auteur: Agasse: Maillard.

Lamarck, J.-B. P. A. de Monet de. 1802b. *Recherches sur l'organisation des corps vivans: et particulièrement sur son origine*. Paris: Chez l'auteur. Maillard,

Lamarck, J.-B. P. A. de Monet de. 1802–1806. Mémoires sur les fossiles des environs de Paris, comprenant la détermination des espèces qui appartiennent aux animaux marins sans vertèbres, et dont la plupart sont figurés dans la collection des vélins du Muséum. *Annales du Muséum national d'histoire naturelle*. Paris: Muséum National d'Histoire Naturelle.

Lamarck, J.-B. P. A. de Monet de. 1809. *Philosophie Zoologique*. Paris: Dentu. Lamarck website www.lamarck.cnrs.fr/index.php?lang=fr.

Lamarck, J.-B. P. A. de Monet de. 1812. *Extrait du cours de zoologie sur les animaux sans vertebres*. Paris: D'Hautel & Gabon. Lamarck website www.lamarck.cnrs.fr/index.php?lang=fr.

Lamarck, J.-B. P. A. de Monet de. 1815–1822. *Histoire Naturelle des animaux sans vertèbres*. Paris: Deterville. Lamarck website www.lamarck.cnrs.fr/index.php?lang=fr.

Lamarck, J.-B. P. A. de Monet de. 1817. Articles from the various volumes of the *Nouveau Dictionnaire d'histoire naturelle*: Faculté, Fonctions organiques, Habitude, Homme, Idée, Irritabilité, Intelligence, Imagination, Instinct, Judgement, Météorologie, Nature. Paris: Deterville.

Lamarck, J.-B. P. A. de Monet de. 1820. *Système analytique des connaissances positives de l'homme*. Paris: Chez l'auteur et Belin.

Lamarck, J.-B. P. A. de Monet de. 1907. Discours d'ouverture: an VIII, an X, an XI et 1806. Paris: *Bulletin scientifique de la France et de la Belgique*.

Lamarck, J.-B. P. A. de Monet de. 1933. *The Lamarck Manuscripts at Harvard*. Eds. W. M. Wheeler and T. Barbour. Cambridge, MA: Harvard University Press.

Lamarck, J.-B. P. A. de Monet de. 1972. *Inédits de Lamarck; d'après les manuscrits conservés à la Bibliothèque centrale du Muséum National d'histoire naturelle de Paris*. Eds. M. Vachon, G. Rousseau and Y. Laissus. Paris: Masson.

Lamarck, J.-B. P. A. de Monet de. 1984 (1809). *Zoological Philosophy*. Trans. by Hugh Elliot. Chicago, IL: University of Chicago Press.

Lamarck, J.-B. P. A. de Monet de. 1990. *Les Introuvables*. Ed. D. Goujet. Paris: Société Française de Systématique.

Laurent, G. 1987. *Paléontologie et Evolution en France de 1800 à 1860. Une histoire des idées de Cuvier et Lamarck à Darwin, Comitté des travaux historiques et scientifiques, Mémoires de la section d'histoire des sciences et des techniques*. Paris: Éditions du CTHS.

Mayr, E. 1972. Lamarck revisited. *Journal of the History of Biology*, *5*: 55–94.

Omodeo, P. 1971. La classification et la phylogénie chez Lamarck. In J. Schiller (ed.), *Colloque international Lamarck* (pp. 11–27). Paris: Blanchard.

Outram, D. 1984. *Georges Cuvier: Vocation, Science, and Authority in Post-Revolutionary France*. Manchester and Dover, NH: Manchester University Press.

Rey, R. 1994. Naissance de la biologie et redistribution des savoirs. *Revue de Synthèse*, *1*(2): 167–197.

Riskin, J. 2002. *Science in the Age of Sensibility: The Sentimental Empiricists of the French Enlightenment*. Chicago, IL: University of Chicago Press.

Roger, J. 1979. Chimie et biologie – des molécules organiques de Buffon a la physico-chimie de Lamarck. *History and Philosophy of the Life Sciences*, *1*: 41–64.

Roger, J. 1995. Lamarck et la Biologie. In *Pour une histoire des sciences à part entiére* (pp. 287–309). Ed. C. Blanckaert. Paris: A. Michel.

Rudwick, M. J. S. 2005. *Bursting the Limits of Time: The Reconstruction of Geohistory in the Age of Revolution*. Chicago, IL: University of Chicago Press.

Rudwick, M. J. S. 2008. *Worlds before Adam: The Reconstruction of Geohistory in the Age of Reform*. Chicago, IL: University of Chicago Press.

Russo R. P. 1981. La notion d'organisation chez Lamarck. In J. Schiller (ed.), *Lamarck et son temps, Lamarck et notre temps* (pp. 111–142). Paris: Vrin.

Schiller J. 1969. Physiologie et classification dans l'œuvre de Lamarck. *Histoire et Biologie*, *2*, 35–57.

Schiller, J. 1971. L'échelle des êtres et la série chez Lamarck. In J. Schiller (ed.), *Colloque International Lamarck* (pp. 87–103). Paris: Blanchard.

Schiller, J. 1981. *La Notion d'organisation dans l'histoire de la biologie*. Paris: Maloine.

Sloan, P. 1995. The gaze of natural history. In C. Fox, R. Porter and R. Wokler (eds.), *Inventing Human Sciences: Eighteenth Century Domains* (pp. 112–151). Berkeley, CA: University of California Press.

Staum, M. 1996. *Minerva's Message: Stabilizing the French Revolution*. Montréal: McGill-Queen's University Press.

Szyfman, L. 1982. *Jean Baptiste Lamarck et son Epoque*. Paris, New York: Mason.

Wagner, P. 2000. An entirely new object of consciousness, of volition, of thought. In L. Daston (ed.), *Biographies of Scientific Objects* (pp. 132–157). Chicago, IL: University of Chicago Press.

Wahrman, D. 2004. *The Making of the Modern Self: Identity and Culture in 18th Century England*. New Haven, CT: Yale University Press.

Wheeler, R. 2000. *The Complexion of Race: Categories of Difference in Eighteenth Century British Culture*. Philadelphia, PA: University of Pennsylvania Press.

Wolfe, C. T. 2014. On the role of Newtonian analogies in eighteenth century life science: Vitalism and provisionally inexplicable explicative devices. In Z. Biener and E. Schliesser (eds.), *Newton and Empiricism* (pp. 223–261). Oxford: Oxford University Press.

Wolfe, C. T. 2016. *Materialism: A Historico-Philosophical Introduction* (Chapters 5–6, pp. 61–86). Dordrecht: Springer.

Postscript 1

A historical proposal around prepositions

Lynn K. Nyhart

To an anglophone historian of post-1800 biology – which may describe the majority of historians of biology – the chapters in this volume might seem quite foreign. They mainly cover the eighteenth century – a century before the one in which "modern" biology is widely supposed to start. They examine historical actors who lived on the Continent, who spoke French and German, not English, and often wrote in Latin. The history of modern biology is dominated by the topics of natural history, Darwinian evolution, genetics, and molecular biology; its historiographical methods in recent years have emphasized the lab and field practices of biology, the geographies of science, and the global circulation of knowledge. To historians operating in this milieu, the present volume's focus on questions of physiology, development and generation, internal and external relations of living things, and philosophy – *so much philosophy* – might be enough to make one put the book right back down on the book table.

This postscript offers an argument for why historians of modern biology (and even philosophers of biology) should keep reading. I want to propose a view of the history of biology in which both the eighteenth century and its philosophizing about vitality are integral to our understanding of "modern" biology, and not just as its historical precursor. My account is necessarily brief and exploratory. It focuses on continuity rather than rupture, on the accumulation of gradual changes that resulted in profound transformation while carrying certain themes forward. It addresses why eighteenth-century continental scholars concerned with questions of life were so focused on philosophy, and how this became so foreign to biology in the succeeding centuries.

I propose that the emergence of modern biology was intimately tied to the narrowing and attenuation of philosophy's ties to biology. This would begin to set up the conditions of possibility for the new philosophy of biology that emerged in the mid-twentieth century, but it also made some of the concerns of the eighteenth century seem foreign. This proposal presents numerous avenues for future research by historians of biology, philosophers of biology who are interested in history, and historical epistemologists.

The very sorts of issues addressed in this volume's chapters provide key clues to this proposed account. But it also queries certain assumptions that seem to me to lie in the intentionally provocative title of this volume: "philosophy of biology before biology." In taking up the provocation, every word in this title deserves some interrogation. Boiled down to its essence, though, the focus of my analysis lies in the relations represented by the prepositions "before" and "of."

"Biology" and "before"

In their broad overview chapter for this volume, Cécilia Bognon-Küss and Charles Wolfe assert that "biology as a science of the functioning and development of living bodies emerged at the beginning of the nineteenth century, integrating methodological or empirical advances in various disciplines, namely physiology, embryology, comparative anatomy, natural history, and medicine." Reinforced by considerations as diverse as the repeated historical reintroduction of the term "biology" in the years around 1800 and Michel Foucault's use of the emergent science of life as a key demarcator between his "classical" and "modern" epistemes,[1] the dating of modern biology, in the sense described by Bognon-Küss and Wolfe, to the period around 1800 is pretty noncontroversial among historians of biology.

But is this assumption really warranted? As William Coleman put it in 1971, "First came the word; a century of incessant activity was needed to create a thriving science" (Coleman 1971, 1). Histories of "biology" as such (as opposed to works on the history of botany, zoology, physiology, anatomy, or other biological subdisciplines) would be a good place to look for evidence of when biology came to be recognized as a thing, as they indicate a consciousness of "biology" as something coherent enough to have had a history. Such histories seem to have arisen first in the mid-nineteenth century, by followers of Auguste Comte, who probably did more than anyone else to popularize the word. In his 1851 *Histoire et systematisation générale de la biologie*, the Comteian Louis Auguste Segond, while mentioning many pre-1800 scientists in passing, championed Xavier Bichat as the crucial figure in the science, reinforcing the "circa 1800" thesis.

As more histories of biology appeared, however, varied opinions arose as to when biology "really" began. Emanuel Rádl, writing in 1905, began his *Geschichte der Biologischen Theorien* with the early modern Aristotelians, but saw the action as picking up with the seventeenth-century mechanists and the vitalists who responded to them.[2] William Locy, in the 1908 *Biology and Its Makers*, dutifully reached back to Vesalius and Harvey but wrote, "In its modern sense, biology did not arise until about 1860, when the nature of protoplasm was first clearly pointed out by Max Schultze" (Locy 1908, 5) – thus aligning himself with a version of biology in which its unity resided in living matter. A lifetime later, the great evolutionary biologist and historian Ernst Mayr wrote in his *Growth of Biological Thought*,

> In the early 1800s, there was really no biology yet, regardless of Lamarck's grandiose scheme and the work of some of the *Naturphilosophen* in Germany. These were only prospectuses of a to-be-created biology. What

> existed was natural history and medical physiology. The unification of biology had to wait for the establishment of evolutionary biology and for the development of such disciplines as cytology.
>
> (Mayr 1982, 108)

Whereas all the above historians treated biology primarily as an intellectual history of key ideas concerning life (and the "makers" of those ideas), in 1988, after a thorough historiographic rehearsal of these and many other historiographic positions, Joseph Caron (1988) argued that the *true* origin of biology lay with T. H. Huxley in England in the 1860s and '70s, because it was under his leadership that something called "biology" was placed into a national school curriculum – thus basing his beginning on institutional/disciplinary criteria.

This brief history of histories of biology suggests we should not assume without further investigation that biology started around 1800, or even earlier – but not later. Instead, we should take seriously the proposition that "biology" itself has been a moving target historically, including the question of when it began. There are different ways to approach this historical multiplicity of biologies. In his postscript to this volume, Philippe Huneman offers a helpful clarifying device: we can distinguish between Biology_1 and Biology_2, where the former is a transhistorical sense of the scientific study of living phenomena and the latter is the historical constellation of concerns that arose and became attached to the word around 1800. Thus, he suggests, we might understand the provocative title of this volume as "philosophy of biology_1 before biology_2." This is extremely useful for analytical purposes, but I think that an even more fluid and capacious approach is warranted.

Our dating loosens up still further if we bring into our scope other quasi-synonyms such as "physiology" and "natural history" – terms that, in some times and places, were used to denote a general science of life and at other times something rather more specific. "Physiology" was an old medical subject, concerning the functioning of the human body and especially its organs, while "natural history" referred to a descriptive and systematizing science that traditionally included minerals as well as animals and plants. Their merger (or proposed merger) under the rubric of "biology" has still only ever been partial, where it has happened, and often uneasy, even as it has generated new insights – from Cuvier's classification, built on functional comparative anatomy, to the conception of modern ecology as an extended form of physiology.

What, then, other than the mere term "biology," justifies making any kind of break at all around 1800? I submit that we would do best not to consider the appearance of the term as representing a sharp break, but rather should view the emergence of modern biology over a longer stretch of time, as part of a much broader transformation of scholarly knowledge, running from the mid-eighteenth century through the mid-nineteenth century, that resulted in the formation of "modern" scientific disciplines.[3]

The key feature of this transformation, I would argue, was the union of natural history with natural philosophy, two aspects of knowledge that in the early modern period were often represented as exercising two different faculties of the mind – memory and reason. Thus for Francis Bacon, and later the *encyclopédistes*, these were separated into the two larger bins of history and philosophy.[4] Natural history's nearest neighbor was civil history; natural philosophy was part of philosophy. The transformation made natural history's nearest neighbor natural philosophy, both parts of the study of nature. This elevated *nature* to a fundamental topic and, by the mid- or late nineteenth century, natural science to a basic division of knowledge, rendering obsolete an organization of knowledge resting on faculties of the mind. The long transition from the mid-eighteenth- to the mid-nineteenth-century organization of scientific knowledge was marked, then, by prodigious efforts and controversies over how best to unite these two aspects of the study of nature, one oriented toward empirical description of the objects of nature and their organization into classificatory systems (natural history), the other oriented toward discovering nature's laws and their causes (natural philosophy). It made the question of how evidence of the empirical world related to the truths of nature a central one, eventually pretty much the only philosophical issue understood to be an intrinsic part of science. The emergence of biology was a signal instance of this transformation, but it was not the only one. It was just one of several new disciplines marked by the union of empirical natural history with natural philosophy.[5]

This little excursus has three points. First, it underscores that the emergence of biology *c.*1800 *was* significant, but that "circa" needs to be extended by about a half-century on either side to show its full significance. Second, to grasp that significance, biology needs to be placed into the broader context of the changing conceptual organization of scientific knowledge. Third, the changing relations of philosophical questions to matters of vitality were key to the consolidation of disciplinary "biology." Attention to these changing relations offers a key way in to understanding what took place in the transformation into modern disciplinarity, which brings me to the rest of the words in our title phrase.

"Biology" (again), "philosophy," and "of"

The chapters in this volume all deal with problems that are simultaneously biological and philosophical. But what sorts of biology and philosophy, and what is their relation? On the biology side, they address problems recognized in the eighteenth century as fundamentally having to do with vitality. The issue of what made living things alive involved a collection of possibilities, areas of theorizing and investigation much more diverse than any straightforward theoretical disputes over preformation and epigenesis or vitalism and mechanism, as Huneman has so succinctly pointed to in his postscript. The research-based chapters in this volume show intense interest among eighteenth-century scientists in the topics of reproduction and development (Schmitt, Sloan, Zammito), transformation (Schmitt, Goy, Gissis), and the relations of organisms to their external environments (Duchesneau, Cheung, Toepfer). Put in the form "How does [reproduction/development/transformation/inside-outside relations] work?" these can be understood not just as topics but as fundamental and persistent *questions* about life. They have continued to animate scholars interested in vitality down to the present day, though they have taken on different vocabularies and approaches to investigation at different times.

One way of integrating the history of biology into a longer-term account, then, would be to consider it in terms of a suite of "questions about vitality," somewhat akin to Bognon-Küss and Wolfe's suggestion (this volume) that we think not about "'biology' in the singular but 'biologies,' which highlights different possible dimensions." To characterize such long-term questions, both philosophers and historians of biology might find it useful to consider the idea of "erotetic organization" (i.e., organization around questions) that Alan Love has found in developmental biology, and to adapt and extend it to biology and its history as a whole. In his studies of modern developmental biology, Love has argued that "developmental biology is organized primarily by stable, broad domains of problems (*problem agendas*)" rather than being organized around theories or models.[6] I would suggest that such problem areas as reproduction, development, transformation, and metabolism could fruitfully be treated as having longstanding question-based agendas as well – *even if*, at some times, they have been organized around particular theories or models. In this expanded view, even as theories have come and gone, the problem agendas encompassing them have persisted. Identifying and tracing these would be a way to bring a strand of contemporary philosophy of biology, with its emphasis on clarifying how scientists actually work, to bear on the *longue durée* history of scholarly concerns with vitality.[7]

This approach, as fruitful as it might be, remains tied to a predominant focus among modern philosophers of biology (shared by many historians of biology) on epistemology and methodology, and does not sufficiently capture the breadth of philosophical engagement with vitality in the eighteenth century. And here is where we must think more about "of" and "philosophy." I would suggest that "philosophy of biology" is inadequate to characterize what was going on in the eighteenth century not exactly because it is anachronistic but because it is anachronistic in a particular way. "Philosophy of biology" implies that biology and philosophy are separate things, and that biology is the thing upon which the critical eye of philosophy is being exercised. Bognon-Küss and Wolfe point out that various terms have been invented to capture slightly different relations between philosophy and biology: Whewell's "philosophy of biology" implied this relation of separation and critique, whereas Comte's "biological philosophy" placed the philosophy within biology, as its theoretical part, and later versions of "biological philosophy" sought to ground philosophy *in* biology. But a yet different relation of biology to philosophy seems, by the light of these chapters, to have been more prominent in the eighteenth century. As Zammito notes in the opening of his chapter, questions about vitality were also an integral *part* of eighteenth-century philosophy, of working out philosophical questions that went far beyond epistemology and methodology.

What happened? Where did the philosophy around vitality go? The history of the emergence of biology as part of the broader emerging disciplinary system of knowledge might be viewed as entailing a kind of piecemeal marginalization or paring away of metaphysical/ontological questions as not part of "the science itself," and their removal, at least in part, to philosophy – a field that was itself gradually becoming leveled, from being the one that governed knowledge to being a discipline itself, if first among equals. To test this proposal, we might try to identify the various ways in which philosophical and biological questions bore on each other, and how those relations changed over time. While this process would necessarily have been patchy and uneven, with different thinkers taking different positions, the chapters in this volume give us glimpses into where we might start in looking for trends, lines of inquiry concerning the shifting relations between philosophy and biology.

Several of these chapters show their historical actors treating fundamentally ontological questions. Stéphane Schmitt's close accounting of the changes in Buffon's early and later roles for molds and organic molecules in reproduction and development concerns, at base, the positing of invisible physical entities in nature – kinds of matter – and their relations. When Phillip R. Sloan writes about Émilie Du Châtelet's interpretation of Leibniz and Christian Wolff's distinction between "substance" and "matter," and the collapsing of that distinction into "vital materialism" by Buffon and Maupertuis, he shows them to be addressing more general questions of ontology and metaphysics. For such questions concerning the basic properties of the stuff of the universe, living things posed a challenge that involved not just the nature of life, but nature, period. Did nature make different kinds of matter – matter with different fundamental attributes? Or could all matter, including that which lived, be reduced ultimately to a single kind of matter such that something else – its combination, organization, or relation to force, gave it the properties of life? When Leibniz, Wolff, and Du Châtelet talked about living matter, they did so within a more general philosophical treatment of matter. Can we trace a change, an autonomizing of these questions as biological ones, such that that larger philosophical framework eventually became reduced to a chapter in a work devoted to a biological topic, a paragraph, a footnote, or nothing?

The broad philosophical and natural-philosophical discussions of the eighteenth century also engaged questions about force. Were the forces involved in sustaining life reducible to those in the rest of nature or not? While in 1800 Jean Senebier had already rejected any appeal to vital forces as "designat[ing] an unknowable entity," which should therefore "be banished from the explanation of phenomena" (Duchesneau, this volume), this position was hardly universal at the time. Only by the late nineteenth century were vital forces largely considered taboo as an assumption or a subject of investigation among working biologists, though this still seems to have varied across national traditions.

We may thus ask: when, how, where, and under what conditions was attention to questions explicitly addressed as "ontological" or "metaphysical" excised from working matters of biology and removed to the domain of philosophy? In which problem areas did they persist? For example, individual development was one feature of life where some biologists continued to explore some kind of vitalism the longest, and the vitalism–mechanism debate remained alive past 1900 – yet, Hans Driesch, the most famous modern vitalist embryologist, having declared his vitalism, left biology for philosophy. Although discussion and debate over vitalism persisted, that discussion was not considered to advance the "work" of biology (Chen 2018).

Questions about purposive organization as a fundamental feature of vitality, whether treated as a fact of nature or in a Kantian sense as merely a regulative idea, have a long history of being considered a legitimate, even defining, part of biology. Georg Toepfer points to the conceptual parallels between Kant's notion that nature involved a "natural system of the purposive relation between different species which exist for each other's sake" and the modern ecosystems concept (though, for Kant, this remained a regulative principle, a model "not given in experience" but only "in teleological reflection"). When did teleology come to be viewed as bad science, "tainting" it, and among whom? How did theorists of biological individuality from the nineteenth century onwards, and systems biologists in the mid-twentieth and since then, work to escape that taint and keep their work scientific?

Toepfer further writes that Kant's ideas led to no direct empirical investigations; "rather, it encouraged romantic speculations on the 'general organism' of nature.... It thus seems that Kant's philosophy of ecology was too early to be effective for science." But some early and even mid-nineteenth century writers who discussed the "general organism" of nature saw themselves as not doing "romantic speculation" but as doing philosophy, and a kind of philosophy that helped explain life.[8] Toepfer's remark thus stimulates a historical question about when and how theorizing was turned into "speculation" in the creation of boundaries between the science of biology and its Others, such as philosophy, speculation, and nonscience, and how such boundary-making was policed and contested.

Similarly, as Ina Goy discusses, Goethe's suggestion that plant metamorphosis was simultaneously symbolic and based on experience was disputed by Schiller, who claimed this could not be an experience but had to be an idea. This *mattered* – at least to Schiller, then a Kantian, and to Goethe, who clearly wanted to say it was both. Within the larger framework of historical investigation I am proposing, Goy and Toepfer's chapters invite this line of questions asking: when did it cease to matter to working biologists what Kant thought? (which parts of Kant's work? which biologists?) What about biologists in France, who cared not about Kant at all? Investigating such questions might help us pinpoint the increasing intellectual separation of biology from philosophy.

Finally, Goy's attention to Goethe's theory of plant metamorphosis, together with Zammito's chapter on the reception of C. F. Wolff's theory of epigenesis, Gissis's work on Lamarck, and Cheung's on inside-outside relations in Cuvier, Hufeland, and Cabanis, set up yet another line of inquiry, about what the treatment of biological *processes* might reveal about the historical relations between biology and philosophy. The Western philosophical tradition is famously better at treating entities and their relations than processes as such, never mind the challenge presented by historicization in the eighteenth-century.[9] Perhaps one reason that development, reproduction, evolutionary transmutation, and metabolic processes were so central to the emergence of disciplinary biology between 1750 and 1850 is because they were not easily encompassed by the terms of philosophy or natural philosophy, and could therefore be more easily asserted as part of a separate field.

Conclusion

The "philosophy of biology before biology" should be provocative for historians of modern biology as well as philosophers of biology. I, at least, have been convinced by this volume to think harder about the historical relations of philosophy to biology and how they have changed. In order to understand the development of modern biology, we cannot simply bracket off philosophy as irrelevant to the "real" biological topics for which we want to know the history. The gradual separation of biology from philosophy *is* the story, or at least a key part of it. Understanding how this separation occurred should offer insights not only into the rise of biology but also into the broader historical transformation in the organization of knowledge from the early eighteenth century to the late nineteenth.

At the same time, my reflections on these chapters and the question of when was "before biology" have reinforced my sense that we should not be fooled by the very existence of the term "biology" into believing it has ever been something very coherent. For the projection of "biology" as a unified science of life has always been a programmatic one, a goal rather than a *fait accompli*, one that has brought together a range of problem areas with distinctive agendas, sometimes masking their diversity. Theories or entities that have been posited as truly providing the unifying key to biology – the cell theory, protoplasm, evolution, the genome – have always privileged certain aspects of vitality over others, and never fully united them all. Yet, a cluster of investigative problems surrounding vitality has involved continuities since at least the eighteenth century. Reproduction, development, transformation, and self-maintenance through interchanges with the external world are highlighted in the chapters of this volume, but we could perhaps add others, such as consciousness and behavior. While the terms have changed in which solutions to these problems have been sought, the biological problems themselves have persisted. In this way, the disunity of problems of life in the eighteenth century may be mirrored in modern biology. It is just that their covering umbrella has changed, from "philosophy" to "biology." And that is, in the end, everything.

Notes

1 Although many historians of biology have objected to Foucault's overstatement and underdocumentation of this epistemic break, it remains powerful.

2 The 1930 English translation, however, began with Darwin and the lead-up to his theory of evolution, thus more or less taking up only Volume 2 of the German original.

3 While "disciplinarity" is sometimes treated in institutional terms, it is at its heart a system of knowledge, and one with a history. This historical formation has itself been considered possibly over, as some consider disciplinarity itself to be on the wane. See Paul Forman (2013).

4 Darnton 1984. Bacon and the *encyclopédistes* hardly exhaust the scope of early modern totalizing systems of knowledge. Yet, even Thomas Hobbes's classification of knowledge in Chapter 9 of *Leviathan* (1651), which Darnton mentions but does not discuss, divides knowledge into two parts: that derived from sense and memory (which he called "history"), and that derived from reasoning (which he called "philosophy").

5 The new Lavoisierian chemistry created a new classification of elements intimately aligned with Lavoisier's theory of what those elements did causally. The new geology of the late eighteenth century brought together the causal (if speculative) science of cosmogony with the descriptive science of mineralogy, to place stratigraphy at its heart. It is worth noting that the same process of uniting the rational and empirical aspects of science – especially the physical sciences – has been viewed as the heart of the seventeenth-century "scientific revolution." As Gary Hatfield has argued, a key part of this was the "separation of experimental mathematical physics from natural philosophy," which, he says, "was consolidated only in the second half of the eighteenth century." See Hatfield 1996, 511.

6 Love 2014, 34.

7 For an example of this with respect to the idea of biological individuality, see Lidgard and Nyhart 2017.

8 Alexander Braun (1851), for example, saw the organism of nature as the highest level of individuality, something that encompassed and explained all lower levels of biological individuals, including the organism and the species. For him, this was a scientific question that he was answering theoretically.

9 Here it is worth recalling that this historicization was evident not only in the historicization of nature but in the very concept of history itself. Reinhart Koselleck's analysis of history famously presents the period 1750–1850 as the "*Sattelzeit*" between the early modern and modern periods, and views the singular "history" emerging out of multiple "histories" as a key emblem of that transformation. For a nice recent analysis, see Fulda 2016.

References

Braun, A. 1851. *Betrachtungen über die Erscheinung der Verjüngung in der Natur*, Leipzig: W. Engelmann.

Caron, J. A. 1988. "Biology" in the life sciences: A historiographical contribution. *History of Science*, *26*: 223–268.

Chen, B. 2018. A non-metaphysical evaluation of vitalism in the early twentieth century. *History and Philosophy of the Life Sciences 40:50*. Online First: https://doi.org/10.1007/s40656-018-0221-2.

Coleman, W. 1977. *Biology in the Nineteenth Century: Problems of Form, Function and Transformation*, Cambridge: Cambridge University Press (original: John Wiley & Sons, 1971).

Darnton, R. 1984. Philosophers trim the tree of knowledge. Chapter Five in *The Great Cat Massacre: And Other Episodes in French Cultural History* (pp. 191–211), New York: Basic.

Forman, P. 2013. On the historical forms of knowledge production and curation: Modernity entailed disciplinarity, postmodernity entails antidisicplinarity. *Osiris 27: Clio Meets Science*: 56–97.

Foucault, M. 1970. *The Order of Things*, London: Tavistock.

Fulda, D. 2016. Sattelzeit. Karriere und Problematik eines kulturwissenschaftlichen Zentralbegriffs. In E. Décultot and D. Fulda (eds.), *Sattelzeit: Historiographiegeschichtliche Revisionen* (pp. 1–16), Berlin: De Gruyter.

Hatfield, G. 1996. Was the Scientific Revolution really a revolution in science? In F. Jamil Ragep and Sally P. Ragep (eds.), *Tradition: Transmission, Transformation* (pp. 489–525), Leiden: E. J. Brill.

Hobbes, T. 1651. *Leviathan*. London: Andrew Crooke.

Lidgard, S. and L. K. Nyhart. 2017. The work of biological individuality: Concepts and contexts. In S. Lidgard and L. K. Nyhart (eds.), *Biological Individuality: Integrating Scientific, Philosophical, and Historical Perspectives* (pp. 17–52), Chicago, IL: University of Chicago Press.

Locy, W. A. 1908. *Biology and Its Makers: With Portraits and Other Illustrations*, New York: Henry Holt.

Love, A. C. 2014. The erotetic organization of developmental biology. In A. Minelli and T. Pradeu (eds.), *Towards a Theory of Development* (pp. 33–55), Oxford: Oxford University Press.

Mayr, E. 1982. *The Growth of Biological Thought*, Cambridge, MA: Harvard University Press.

Rádl, E. 1905–1908. *Geschichte der biologischen Theorien*, 2 vols., Leipzig: Engelmann.

Segond, L. A. 1851. *Histoire et systématisation générale de la biologie*, Paris: J. B. Baillière.

Postcript 2

Philosophy after philosophy of biology before biology

Philippe Huneman

Besides the fact that the two words start with a B, biology and the baroque have something in common: they both have at the same time a general ahistorical meaning and a precise, historically situated, meaning. 'Baroque' is a school of fine arts that was developed in the seventeenth century in opposition to post-Renaissance classicism. It's also a family of styles that cultivate apparent disorder, sophisticated proliferations, rich ornaments, lack of symmetry, etc. – and is most easily captured by being opposed to classicism as an eternal category. In this sense, each period had its 'baroque', and 'classic vs baroque' constitutes an interpretative scheme that can be applied to any of the arts, at any period: Chaplin vs Welles, Cartier-Bresson vs William Klein, Erroll Garner vs Thelonious Monk, Schubert vs Schumann, or even Carver vs Pynchon.

Similarly, biology is very generally understood as any scientific reflection on living phenomena – so there is a biology of Aristotle, of Galen, of Harvey, etc. But many historians of science privilege another meaning, according to which, as Foucault famously said, biology didn't exist before the nineteenth century, not because no scientist was interested in studying living phenomena but rather because an autonomous science devoted to the functioning and developments of living beings as alive wasn't established before the end of the eighteenth century. Natural history, physiology, and medicine were extant bodies of knowledge dealing indeed with life and living beings, the proper structures of which were inherited from Hippocrates, Galen, and Aristotle, but they did not together constitute a unified specific science that would have singled out 'life' and its many declinations as its single object. Let's call those two meanings, the ahistorical and the historically situated, respectively biology$_1$ and biology$_2$.

This claim on the recent birth of 'biology' has been partially challenged by many scholars, yet it's hardly objectionable that some break occurred in the scientific approach to life around the turn of the nineteenth century. But the project of the present volume can easily be understood once one sees in 'philosophy of biology' two successive occurrences of the two meanings of the term, something like 'philosophy of biology$_1$ before biology$_2$'. So the object of the contributors here seems to be a philosophical reflection on how one should conceive of and scientifically study living phenomena before there was one autonomous and unified scientific discipline dedicated to the study of living phenomena. After all, nobody would object to a 'philosophy of the baroque before the baroque', which would address, for instance, Rabelais, Villon, Breughel, or even Shakespeare, instead of the overtly baroque artists who historically vindicated the label.

Besides this duality, the scholar of biology necessarily meets two orthogonal conceptual dualities that informed biological thought since at least the Renaissance, namely 'mechanism vs vitalism' and 'preformationism vs epigeneticism'; and, exactly like the notion of 'biology', both also allow of a historical and a ahistorical reading. Regarding the latter distinction, preformationism and epigenesis were two distinct stances taken towards the problem of generation in the seventeenth and eighteenth centuries – preformationism being first initiated by philosophers like Leibniz and Malebranche, and scientists equipped with microscopes like Swammerdam, Leeuwenhoek, and Malpighi; epigenesis having been in turn vindicated by authors like Maupertuis, Buffon, Wolff, or Blumenbach. There is a huge body of scholarship on this controversy, and the received view is that epigenesis somehow won, with Von Baer's groundbreaking 1828 treatise on development (*Entwicklungsgeschichte den Thieren*). At least nobody now believes in the pre-existence of germs, which was the particularly outdated and extreme form of preformationism held by Malebranche (the 'Russian dolls' story about generation, so to say). However, there is another, less historical, sense of those two terms: preformationism is an emphasis on the fact that in any development some form is presupposed and initially there, inherited from the parents, while epigeneticism stresses the indispensable causal role of possibly physical and chemical forces that bring about the adult organism. In this sense, those two terms are rather two poles of embryogenetic thinking, and they are still with us: the 'genetic programme', for instance, is obviously a preformationist notion, while self-organization is one current avatar of epigeneticism.

As to mechanism vs vitalism, another major structuring distinction for reflections on the living, but wholly decoupled from the former, things are very similar. There have been mechanicists, like the supporters of Cartesianism in physiology, like Boerhaave, who called themselves 'iatromechanists' – and authors who vindicated vitalism and who built vitalist schools of thought, one of the most famous being the Montpellier school of vitalist medicine in the eighteenth century.

But here too it is easy to see that mechanism, understood as the insistence on the constitutive role of physical processes in any biological function and biological development, is a key aspect of biological thinking, and could not reasonably be seen as something that will one day go away. Vitalism, in turn, taken in the most minimalist sense, is the view that, no matter the physico-chemical determinants of vital phenomena, there is something in these phenomena that justifies that they are the object of a somewhat autonomous science – a point that couldn't be cast aside easily as long as a discipline like biology exists. In the second sense, then, neither preformationism nor epigeneticism, and neither mechanicism nor vitalism, could be wrong; they are stances, perspectives towards what's alive, and within biology$_1$ – in the sense of an eternal kind of science – they seem to be inseparable. While Canguilhem was among the first to expose the recurrences of those two couples of notions, Holton's idea of 'themata' nicely provides a frame to make sense of such a trans-historical existence of an idea.

Thus, by 'philosophy of biology before biology' I take it that authors of this volume investigated what, among the philosophical theorizations and interrogations of biology$_1$, at the times of the gestation of biology$_2$, has been somehow incorporated in what should become the philosophy of biology$_2$.

As expected, those two themes, mechanism/vitalism and preformationism/epigenesis, receive novel and fruitful treatments here, for instance in chapters by Sloan, Zammito, Schmitt, and Duchesneau. But, rather than discussing these treatments, I'll present in what follows a few reflections of my own prompted by the materials exposed in the volume, and regarding other topics, distinctions, or simply conceptual issues pertaining to biology$_1$. I will conclude with some comments on the relationship between history and philosophy of biology as they appear throughout this book. Thus the present postscript is less an appraisal or a recollection of the book than a set of 'free associations' based on the reading of such an innovative and inspiring collection.

(Philosophy of) biology then and now

Reading this book did not provide me with a novel narrative about the emergence of biology, the rise of an autonomous science, or even the intertwining of new disciplines that displaced natural history, medicine or physiology. The volume revisits some rather well-studied figures, like Lamarck, Bonnet, Cabanis, Blumenbach, and Wolff; more marginal scientists are also put to the foreground such as Senebier, Mme Du Châtelet, Hufeland, and Tetens. This sole fact allows one to reconsider many of the major narratives proposed of the intellectual events through which biology$_2$ emerged as a science – the triumph of some form of organicism over both mechanism and vitalism, the rise of experimental physiology and clinical medicine that overcame the purely abstract ontological medicine and 'design-centred' physiology (e.g. Cunningham 2003), the slow recognition of the epistemological autonomy of living phenomena, etc. An appreciable effect of this collection of chapters that often focus on some 'minor' figures thus consists in creating a kind of excentration with respect to biology$_2$.

As it stands in this volume, the usual intellectual heroes – Haller, Montpellier vitalists, Kant, Blumenbach, Lamarck, Cuvier, Goethe, Wolff, Bonnet, Buffon, etc. – were not isolated players but part of a galaxy of very general intellectual innovations and shifts, which concerned not only those who directly dealt with animals or plants – physicians, taxonomical botanists, experimental physiologists, etc. – but also those who worked on the foundations of physics, the reform of metaphysics, rational or empirical psychology, or the elaboration of mathematical calculus. And their debates, discussions, exchanges, and misunderstandings were all part of what shaped the philosophy of biology$_1$ that is investigated here. Seen from this perspective, the chapter by Gissis on Lamarck contributes to the same message: there is more in Lamarck – and Lamarck as a major biologist – than the *Flore française*, the classification of invertebrates, the *Hydrogéologie* and 'Lamarckian inheritance'; there is a set of elements proper to distinct worldviews which pertains not only to biology, but to what will later on become social sciences. Thus, both Lamarck's transformism and Lamarck's critique of vitalism and 'emergent materialism' (as Baertschi (1992) some time ago characterized it) are pieces in this interplay between politics, reflections on living things and the novel, triumphant science of nature whose rising Newtonian figures were Lagrange and Laplace. The very recent hype from which the scientific idea of epigenetic inheritance has benefited – idea defended by Jablonka and Lamb (2005), or Danchin *et al.* (2011) among many others – nicely echoes this chapter on Lamarck: while many authors consider epigenetic inheritance as the revival of Lamarckian inheritance or at least as a Lamarckian idea (the 'Lamarckian dimension', as Jablonka and Lamb labelled it), and while many social scientists see it as a bridge between the biological and the social in the human species (e.g. Meloni 2016) – since socially determined facts may impinge on biological descent through epigenetic inheritance – it' s enjoyable to see the great extent to which the original idea of Lamarckian inheritance by Lamarck was already impacted by concerns about society and social science.

Thus, the present work, while it doesn't offer us a new narrative, does provide us with resources to think through issues that were the subject of dispute both then and now.

Information and metabolism

As recalled above, and as Holton and Canguilhem made clear by distinct means, biology$_1$ is torn between two sets of rival stances: mechanism vs vitalism, and preformation vs epigenesis. Those are rival stances in the sense that they map a disagreement about how life and generation should be understood, and what is ontologically distinctive of life. Yet another conceptual distinction is crucial to appreciate what is epistemologically at stake in the practice of biology, namely the 'form vs function' distinction, as forcefully explored by E. S. Russell (1916).

This distinction is not of the same nature as the two others, since it's not likely to be formulated as a pair of opposed theses about the nature of life. Mechanists reject vitalism and vice versa, but biologists focusing on forms, such as Geoffroy Saint-Hilaire, Goethe, or Pere Alberch, do not reject the idea that organisms have functions; they just think that it's epistemologically less relevant. And this holds inversely for biologists, such as Cuvier, Huxley, or current-day adaptationists in behavioural ecology, for instance, who stress function. The opposition may take the form of an open controversy, as was the case with Geoffroy Saint-Hilaire and Cuvier's famous debate at the Museum d'Histoire Naturelle de Paris in 1830, but such open confrontation is just an expression of the difference between form biology and function biology, not its nature. Of course, many biologists – starting with Aristotle – were located in the middle of this continuum between form biology and function biology; nevertheless, considering this distinction is still useful to make sense of debates, epistemic disagreements, or methodological differences within the sciences of life since Aristotle.

This conceptual distinction is therefore logically distinct from the two pairs of concepts that I mentioned above: it does not express two rival theses and it has only a ahistorical sense, not two senses (a historical and an ahistorical one).

But a thread appears across many contributions in this volume: the acknowledgement that form/function is *not the only structuring conceptual distinction in biology*, it's not the only one of its kind, so to say; another one actually surfaces here, which, tentatively, I'll label *information vs metabolism*.

Granted, this may be a familiar distinction among those who currently work on the origins of life where replication-first accounts, focusing on the emergence of genes, genetic code, are contrasted with metabolism-first accounts, focusing on the evolution of cell metabolism (e.g. Fry 2002). But philosophical questioning before biology$_2$ shows that a very general tension between two poles was already structuring the debates long ago: on the one hand, life is characterized by the efficiency or existence of something that somehow constrains organisms both in their repertoire of behaviours and in their emergence; on the other hand, life, whatever it turns out to be, is characterized by a material process through which it both maintains itself and it turns the non-living into living. While those two poles have been captured technically by particular concepts, and the major concepts used nowadays are metabolism for the latter and genes (or epigenetic etc.) for the former, it's plausible to view such poles as '*information*' and '*metabolism*'.

That is indeed something that is established by Schmitt here, regarding Buffon's conceptual trajectory: his fluctuations about the privileged role of 'organic molecules' – which should be understood as displaying some specific abilities – and 'interior molds' may testify to the presence of those two structuring poles within his views of generation. To cite Schmitt about this bipolarity, 'on the one hand, "organic molecules", the components of the organic matter, endowed with particular properties; and, on the other hand, "interior molds" that supposedly organized the molecules into organisms in the course of the embryonic development'. But the complex shifts that constitute the elaboration of epigenesist thinking in early nineteenth-century German life sciences, as described here by Zammito, studying the reception of Wolff's descriptive embryology, also indicate such a polarity. While obviously the concept of a *Bildungstrieb* and its recognition by Blumenbach as what drives embryogenesis – a long and complex recognition process whose steps are convincingly explored by Zammito – pertains to the long-run history of epigenesis, one should notice that the scope of the *Bildungstrieb* is somehow fluctuating within Blumenbach's thinking. It obviously contains what we could call some 'information' (about the species type), but Blumenbach also ends up by including the force of nutrition within this *Bildungstrieb*. Yet, nutrition is a process through which metabolism is carried on at all levels; therefore, one could also read, within the history of the constitution of epigenetic thought, a side story about the subsumption of metabolism within information: a subsumption which echoes, within a different conceptual and epistemic environment (namely, discussions of Kant's purposiveness, controversy with Haller, etc.), the fluctuations in Buffon's thought analysed by Schmitt. And to some extent the story told by Cheung about the role of the processes achieving a separation within the inner and the outer, conceptualized so differently by Cuvier the comparative anatomist and the physicians Cabanis and Hufeland, could be understood as an episode within the integration of the pole 'metabolism' into this nascent biology$_2$, as well as an initial recognition of its irreducibility. Granted, Cuvier, who is examined in Cheung's inquiry, is supposed to constitute a major breakthrough in the tradition of 'function biology', as Russell and later Amundson (2005) defined it; but this chapter indicates that, concerning the other axis, namely metabolism/information, his comparative anatomy also participated in the elaboration of a set of accounts that became definitional of the metabolism-oriented tradition of biology.

Notice that one shouldn't confuse the present distinction between information and metabolism with the form/function distinction, even though information may sound like form. Genes are information, but they are not 'form' in the sense of the form/function distinction. To take a contemporary example, Pigliucci (2007), in the very precise context of the discussions about a possible extension of the classical theory evolution (i.e. the so-called modern synthesis) towards new concepts (epigenetic inheritance, directed variation, etc.), says that Darwinian biology was mostly focused on *genes* (hence, information), and forgot about *form* (an interest instantiated by the tradition of D'Arcy Thomson, Pere Alberch, Brian Hall, etc.) – which presupposes of course that form and information are two different things. And, even though *metabolism* is about some processes, and that physiological *functions* at the level of the cell and the organisms contribute to metabolism, it's wrong to conflate it with 'function' in general, since nothing in the essence of the concept of 'function' ties it to the transformation of external matter into matter akin to the organisms' matter, or to the establishment and preservation of an organism's boundaries.

Here, then, we have a second structuring pair of concepts and the philosophy of biology before biology helps us to make sense of its relevance for biological thought. Inversely, by learning about the persistent and underlying role of this distinction in discussions that preceded the constitution of biology$_2$, we see how in fact 'philosophy of biology$_1$' can be relevant for biology$_2$.

Biology and chemistry

Metabolism arguably concerns biological organization. As Duchesneau's chapter on Senebier, Zammito's on Blumenbach, and Gissis's on Lamarck make clear, a major concern in the period of this philosophy of biology$_1$ that accompanied the rise of biology$_2$ was the establishment of an epistemic strategy to capture *laws of vital organization.* A section of the present volume indeed is devoted to 'Organization'. And what Duchesneau writes about Senebier could be applied to many of the figures encountered in this book: Mme Du Châtelet, Lamarck, Blumenbach, Buffon, or Cuvier. Quoting Duchesneau here:

> The challenge is to infer from these facts causes and laws that would own the highest epistemic probability and truly account for the order of phenomena concerned. In the context of Senebier's researches, the aim is thus to start from accessible experimental data and discover the causes and laws that would explain the emergence and order of production of such complex organic and vital phenomena.

Of course, this idea has already been explored by many scholars. But two interesting threads are developed in a novel and fruitful way in the present collection. The first one is about the differences in philosophical understanding of the notion of 'laws', to which I will return later; the second concerns the relation between chemistry and biology, on which I'll elaborate now.

Many chapters indicate that the relation between this 'organization' deemed to be proper to life, and chemistry, as a set of phenomena that might also obey laws distinct from the laws of physics (even though, unlike biology, those phenomena don't seem to be proper to one single ontological kind), is perplexing.

Once again, I quote Duchesneau on Senebier:

> The demonstrative strategy set up consists on the one hand in determining experimentally: (1) the sources of supply of the CO_2 obtained by dissolution in water: (2) the absorption of the latter by the peripheral organs: roots and leaves; (3) the modalities of its decomposition by the action of light without emission of caloric; (4) the corresponding exhalation of oxygen by the leaves; and (5) the combination of carbon along specific affinities in the various solid and liquid components of the vegetal organism. In all cases, experiments are devised so as to fit the diversification and variation of parameters and facilitate quantifying the processes and products resulting from vegetative activity.

The general register under which those experimental questions should be listed is mostly chemistry; 'vegetative activity', the object of plant physiology according to Senebier, is mostly lumped together with a class of chemical phenomena, and should be addressed through chemical experiments.

As Gissis acknowledges in her chapter on Lamarck, Lavoisier's work on respiration was foundational for such approaches. Indeed, this work paved the way for many attempts to experimentally design, by employing chemical categories, a scientific approach to vital functions and functioning. In the case of Senebier, the chemical approaches to plant's respiration, photosynthesis (not yet known), string of CO_2, etc – especially by Ingenhousz in the middle of the century – were bases for his own enterprise of systematic physiology.

One may here reflect along the following lines. Focusing on chemistry makes salient a paradox that is often overlooked by the histories of the emergence of biology, which are often concerned by the process through which biology became independent from 'physico-chemistry' or 'the science of matter'. The common narratives about the rise of scientific mechanism against spiritualist or religious vitalism, or, inversely, of the vitalist alternative to a mechanical undermining of the specificities of life – these narratives lump together physics and chemistry (both being opposed to biology).

However, some of the situations explored here indicate that, while biology emerged as autonomous, it was at the same time embedded within some forms of chemistry (Bognon-Küss, forthcoming). The duality of mechanism/vitalism, whether in a crude or in a sophisticated form, always disregards this intertwinement of biology and chemistry. Such disregard is surely in part the effect of historical distance. After all, for Kant it was obvious that chemistry and biology share many things, and he opposed them as a whole to 'physics', stating in the *Metaphysical Foundations of Natural Science* that neither biology nor chemistry are sciences since they are not founded on a priori mathematical principles.

And at the same time this intertwinement is something very old, predating biology$_2$. Ingenhousz, like Stahl, was a physician and a chemist. Moreover, for Stahl one of the issues raised by living phenomena is the determination of their proper chemistry – the kind of 'mixt' proper to life, and accounting for the possibility of the resistance of organisms to common chemical phenomena.

Thus, besides any commitment to vitalism, mechanism, materialism or whatever, the fact is that nascent biology$_2$ required, as much as being autonomous from 'physico-chemistry', an embedding within a research programme targeting the 'chemistry of life', something that will later on become organic chemistry and then biochemistry.

This becomes all the more clear in the middle of the nineteenth century, precisely when Liebig and others designed a programme for organic chemistry, which in turn became autonomous from biology, and gave rise to biochemistry. But, in the period where philosophy of biology before biology was discussed, this intertwinement of life and chemistry was epistemologically something salient. Even though nowadays some scholars of course acknowledge this – especially as a preamble to several histories of organic chemistry or biochemistry – this book explores it with great force. In fact, the editors in their chapter call for more attention to such interplay between chemistry/biology when it comes to recognizing the specificity of biological organization and set up research programmes to address it. I just wished to explore this topic slightly further once the book has been read.

Here, the mere consideration of terms and their grammatical affinities can help us to further explore the matter.

Organism is the focus of biology$_2$ as a science – as has been forcefully investigated – and Kant's philosophy legitimately attracted some attention because it contains a major expression of this key role of the concept of organism. *Organization* is a concern for biologists and natural philosophers – many of them considered in this book – who wondered about the 'laws of organization'. *Organization* may extend beyond organisms, as is explored by Toepfer's chapter, which addresses Kant's theory of external purposiveness, rather than Kant's view of organism – that is, his views of the organization of the nature of organisms and, then, the views that evolved from these conceptions.

An 'organism' is '*organized*' – as Cheung (2006) established, 'organism' was initially introduced by authors such as Leibniz to mean a kind of organization, before it came to denote the substance organized in this way, namely an individual animal or plant. The organization of an organism is of course its being constituted by organs, which are in turn organized into tissues, etc. The nestedness of organisms is something that has been familiar to biology$_1$ since Leibniz, and a trend in the history of biology consists of always zooming in on this hierarchy and unravelling lower levels of organization: tissues (Bichat), then cells (Schwann), then chromosomes, then genes, then DNA, etc.

But things other than 'organisms' can be organized. It makes sense to talk of the organization of a molecule, provided that we understand organization as 'structure', which is a possible way of conceiving of it. *Organic* molecules are organized, then (and, after all, a major event in the foundation of modern chemistry is the unravelling of the structure of the molecule of benzene by Kékulé, and the formulation of the notion of chemical structure). Organic chemistry is the science of the organization of those molecules, made up of atoms of C, H, O, and N.

Thus, 'organization' can surely be predicated *above* the level of the organism – the organization of the whole of living nature, as Kant conceived of it and as Toepfer in his contribution explicates it and shows to be connected to German romantic ecology – as well as *below* the level of the organism, and this is the 'organization' discussed by organic chemistry.

Hence, the mere word 'organization' can easily shift from 'organisms' to 'organic'. Those shifts are lurking behind Buffon's fluctuations as described by Schmitt in his chapter. This family of words therefore constitutes a mediator between biology and physics, and it's not absurd to think of it as the semantic locus within which the intertwining of biology and chemistry, which accompanied the emergence of biology$_2$, could take place.

And then, at the higher level, a strange question was of interest here: is organized nature, as a whole, alive or dead? As Stéphane Schmitt recalls, Buffon finally claimed that all brute matter is a residue of living matter – especially inorganic molecules, which have been produced by living bodies. This apparently strange view was not unique. Zammito points out that Blumenbach's *Bildungstrieb* was eventually incorporated within biology as a property of organisms, but that initially he ascribed it to the whole of nature, conceiving of it as the inbuilt tendency of nature to build forms. Thus it is troubling that the scope of biology was rather undetermined – spanning both what will later be organic chemistry and ecology – in the period where biology$_2$ emerged.

Another merit of the collection consists in putting in parallel several conceptual itineraries, such as Buffon's, Blumenbach's, Senebier's, or Lamarck's. It strikingly shows how none of these itineraries was direct and linear: all of the authors found and expressed their final claims after having defended various positions, like Buffon on 'interior molds', Blumenbach on the *Bildungstrieb*, etc. Their itineraries sometimes navigate the wide conceptual field that yields the equivocacies of the family of words like organism, organization, and organic, which I surveyed here.

Biology and metaphysics

'Laws' is a very equivocal word. The 'laws' of organization can mean many things according to the philosophical account of 'laws' one holds. And in this book the possible philosophical *rationale* for the specific project of explicating the laws of organization are multiple.

Which takes me to my fourth observation, namely that a strength of this book consists in mixing biologists like Senebier, Hufeland, or Spallanzani – and philosophical contemporaries such as Kant, Leibniz, and Newton, among others.

Briefly said, in this perspective such a study reminds us that attaining a firm ground for biology$_2$ implies not the commitment to one philosophy but a set of various commitments and ontological claims that refer to many philosophical positions. Scientific disagreements on life, its development, and its functioning, emerge along the structures of the problematic field I sketched above, but, to be complete, I should add that they also range along a gallery of two or three major philosophical – namely, metaphysical and epistemological – sets of commitments.

Checking the book, I see that Kant, Leibniz, and after him Newton are cited throughout (granted, one chapter is partly on Kant and another one touches on Leibniz via Du Châtelet). This is not surprising. Kant is a major philosophical reference, and the recent rise in interest for his theories among philosophers of biology may explain his prominence in this volume. Newton and Leibniz as explicit references of the biologists of the times are naturally expected here. One major investigation of at least German biology is the recent book by John Zammito (2018), the *Gestation of German Biology*; in this major work, Zammito's interpretative hypothesis is that Newton and Newtonianism, understood as a rather experimental research programme – leaving aside the mathematical component of Newtonianism – provided the elements of the theoretical framework in which biology emerged. This is not a shocking view: anyone familiar to life sciences at this period knows the extent to which Newtonian science was a constant reference for life scientists at the times, and Zammito emphasizes that his claim is about German biology, and not biology as such (even though the French vitalists play a major role in his narrative, given that they were extensively read by Germans after in turn having been inspired by some of them). In the present book, this approach is, if not challenged, at least complemented by a general insistence on the role of *Leibniz* for philosophers discussing biology$_1$ at the times biology$_2$ was emerging. Schmitt places Buffon's intellectual journey under the sign of Leibniz and concludes that his final view regarding molds and organic molecules takes him very close to Leibniz. Sloan, in his chapter, minutely explores Du Châtelet's work in the light of biology, and reconstitutes the network of her influence; in turn, her programme was mostly concerned with bringing Leibniz's science into modern physics. Sloan's inquiry into the role played by Mme Du Châtelet in the emergence of biology$_2$, and his diagnostic that a parting of the ways between her Leibnizianism and more Newtonian views operated there, is very illuminating for anyone who wants to identify the philosophical commitments made by the promoters of biology$_2$.

Thus, the field of biology$_1$ just before biology$_2$ appeared articulated or torn between three major references: Leibniz, Newton, and Kant. (One will notice that both Leibniz and Kant discussed Newtonian physics.)

Let's elaborate a bit on these bases. The relationship between philosophy and science is a complex matter, and has evolved massively throughout history. While we mostly regard Bonnet, Spallanzani, and Du Châtelet, and even Newton, as scientists, they were considered 'natural philosophers' in their own time and the distinction we make between them and, say, Leibniz or Malebranche, is retrospectively projected on them. It could be argued that the difference between science and philosophy as we make use of it started to emerge with Kantianism; it could also be argued that this difference is not shaped in the same way when it concerns a well-established scientific field like physics and an emerging field like biology or chemistry – a Kuhnian would here appeal to the pre-paradigm/post-paradigm difference but I won't commit to any Kuhnian account here.

The reference to philosophers like Kant and Leibniz in discussions surrounding the emergence of biology$_2$ is thereby not to be taken exactly in the same way as the reference to those who are, to us, naturalists – Spallanzani, Haller, Bonnet. What is in general appealed to is not claims about biology proper – even though Kant or Leibniz did make such claims – but ontological and epistemological stances supposed to be illuminative for the practice of biology. As Sloan writes, Leibniz's metaphysics is referred to, through Du Châtelet, by naturalists interested in living phenomena, mostly because it proposes an ontological theory of autonomous substances.

Yet, something interesting in such references is that often the 'natural philosophers' – or, for us, scientists – don't share the problem agenda of the philosophers; they may even not bother with the metaphysical subtleties that are the bulk of the philosopher's reasoning. The relation between Blumenbach and Kant, as investigated in Zammito's chapter, may be seen as paradigmatic of this kind of appeal or reference. Blumenbach wasn't really interested in Kant's major advances: the theory of regulative principles of judgement, the antinomy of teleological judgement, etc. Mostly, the Kantian notion of purposiveness, taken independently from the transcendental philosophy within which, for Kant, it could only be meaningful, could back up Blumenbach's account of mildly epigenesist embryogenesis, which was his major interest. Zammito speaks of 'creative misunderstanding' here (after Richards 2000), and I'd suggest that this category may cast a light on the reference from 'scientists' to 'philosophers' at the times of the constitution of biology$_2$.

But that is not all there is to the presence of Leibniz and Kant in those pages. Even though the scientists themselves were not interested in the philosophical justification of the ontological or epistemological claims they wanted to borrow, it's plausible that subscribing to Leibniz's or Kant's philosophy implied further commitments than just borrowing one or two claims one thinks to be true in the context of one's scientific activity.

The last point I wish to make here is thus that those philosophical references are also shaping ways of doing science (and don't just back up scientific claims, such as mild epigenesis). Recent philosophers (Bird 2007, Ladyman and Ross 2009) speak of 'metaphysics of science' in order to name the philosophical activity that consists of developing philosophical answers to metaphysical issues – such as causation, laws, time, etc. – based on (or at least concerned with) our best current science. A possible reading of some of the accounts of 'philosophy of biology before biology' developed in those pages would then go along the following lines: Leibniz or Kant elaborated systematic and complete metaphysical frameworks in which some biological concepts and some methodological formulations could be developed; for biology$_2$, they defined competing or alternative 'metaphysics of science'. Thus, among the scientists whose joint activities constituted biology$_2$, besides the solidarity of their actions (for instance, embryologists like Wolff and physiologists like Senebier looked for laws of organization – as functioning, or as developing organization – and hypothesized irreducible forces or properties in a seemingly Newtonian manner), divergences in metaphysical frameworks could be effective.

In other words, even if biology$_2$ is somehow a unified, structured (into disciplines), and autonomous science, as scholarship has established, at another level it could be viewed as divided into distinct metaphysical and philosophical frameworks/bases/foundations. Kant and Leibniz here would be major labels for those metaphysical foundations. In this sense, the gap between Bonnet or Buffon on the one hand, and Blumenbach or Von Baer on the other, does not merely concern the role of experiments, and is not merely determined by the preformationism/epigenesis split, but relies on the difference between a Leibnizian ontology of nested substances and a Kantian understanding of causation and mechanism.

I'd say here that, beside some specific historical and theoretical topics handled by the particular chapters, the book as a whole suggests the idea that, behind the distinction of disciplines and methodologies among scholars interested in living phenomena at the period, behind the unity of emerging biology$_2$, stands a radical distinction between competing metaphysical commitments, mostly involving Kant and Leibniz as key figures.

If one adopts this suggestion, one sees how the philosophy of biology before biology could be relevant to current philosophy of biology. Given this level of metaphysical signification that somehow coexists with the scientific unity of biology, one could still question the metaphysical underpinnings of current biological projects. This is not wholly new: some authors have emphasized that Kant's view of organisms shapes to some extent the 'developmentalist' stance on evolution (Gilbert and Sarkar 2000). Others have suggested that Leibniz's conception of individuality as nested is realized by the current Darwinian view of individuality as the result of a multilevel selection process (Nachtomy *et al.* 2002). The current volume suggests to some extent that this doesn't just concern two disputed and isolated theses, but that it's quite general and it's about the fact that distinct philosophies have coexisted within biology$_2$ since its inception, and that differences between Leibnizian and Kantian metaphysical underpinnings still pervade many current research programmes, especially when it comes to issues like individuality, organism, causation, and teleology. In this perspective, while there is an Aristotelian biology – which some scholars now intend to renew in the context of current developmental biology and Evo-Devo, for instance Walsh (2015) – or a Cartesian biology (which seems to be outdated), a peculiarity of biology$_2$ is that it has been elaborated in a landscape shaped by the figures of Kant and Leibniz. Even though most of the biology$_2$, methodologically speaking, was indebted to Newton, its pathways are still epistemologically or ontologically determined by such a major opposition between Leibniz and Kant.

This implies that some philosophical questions currently raised, for instance about 'reductionism', or about biological individuality, may be too grossly formulated. If there have been more Leibnizian and more Kantian provinces of biology since the nineteenth century, at least from a metaphysical viewpoint, it means that questioning 'reductionism' in biology as some compact entity risks being too coarse-grained. Equally, philosophizing about biological individuality, as is often done in recent philosophy of biology, risks a major misunderstanding if it is assumed that there is one concept of individuality used by biologists, since Kant-leaning and Leibniz-leaning biologies would differ on this precise issue – the status of 'individuality' being a point of crucial divergence between Kant and Leibniz, the latter conceiving of infinitely nested individuals as characteristic of biology, while the former explicitly rejected infinite nestedness and insisted on his double criteria for natural purposiveness (§65 of the *Critique of Judgment*) as definitional of biological individuals within our knowledge. And when, as often, philosophers conclude their reflections on biological individuality by acknowledging a plurality of concepts depending on fields (e.g. Godfrey-Smith 2013), this pluralism might very well be enriched by recognizing the metaphysical pluralism or cleavage that, according to an interpretation of the present book, was constitutive of the emergence of biology$_2$ and still possibly pervades biology.

This hypothesis assumes that a focus on vitalism vs mechanism is not the only or the best way to address the metaphysics of biology. While it has been previously tempting to question whether Kant is a vitalist or a mechanist, and the same for Leibniz, a more fruitful approach would consist of taking Leibniz's and Kant's philosophy as irreducible wholes without invoking the labels 'vitalism/mechanicism' (which would be misleading in this perspective) and then questioning how the distinct research programmes in biology behave regarding those wholes, namely regarding Kant's stance on mechanism, Leibniz's stance on nested individuals, etc.

Philosophy and history of biology

This last suggestion leads me to a final remark. How philosophy of science should relate to history of science is an open question, about which philosophers have thought and discussed extensively since the 70s – with a revival of interest for this issue in the last decade. Here, in the perspective provided by this book, it seems that history of biology may provide us, at least, with some insights into the metaphysical underpinning of biology, or, at least, some research programmes, traditions, and science in biology. Granted, an exact knowledge of Senebier's physiology may seem of no use for the philosophy of science; and examples like this support the thesis that, in the end, philosophy of science and history of science are two distinct and non-overlapping scholarly activities. However, knowing how Senebier conceived of his project of plant physiology in relation with philosophical views about laws and explanation, and how these views relate to some competing philosophical frameworks that were involved in the formulation of the conceptual threads that concurred into defining biology as a science, is arguably very relevant for philosophy of biology, especially to the extent that the conceptual space in which biology evolves has partly been constructed when such biology$_2$ emerged, or, at least, that current biology still encompasses some research programmes (like developmental biology, or the project of a science of the organization of organisms, as Toepfer investigated here) that arose during such scientific breakthroughs.

In order to characterize this presence of Kant and Leibniz as major informant philosophical figures for philosophy of biology before biology – to which, very possibly, later on Hegel should be added – I used several distinct words. I spoke of 'metaphysical foundations', 'frameworks', 'underpinnings', 'horizon'; the hesitations here reflect that such a presence, which this book immensely contributes to detect, is not of an unequivocal nature. On the contrary, defining this as a metaphysical presence within biology is – if one accepts the last suggestion made here – an open question, which undoubtedly pertains to philosophy of biology, no matter the stance one takes regarding the relationship between philosophy and history.

Acknowledgements

I am grateful to the editors for their careful reading and comments. Insightful critiques and suggestions by Boris Demarest and an anonymous reviewer greatly improved the manuscript.

References

Amundson, R. 2005. *The Changing Role of the Embryo in Evolutionary Thought: Roots of Evo-Devo*, Cambridge: Cambridge University Press.

Baertschi, B. 1992. *Les rapports de l'âme et du corps; Descartes, Diderot et Maine de Biran*, Paris: Vrin.

Bird, A. 2007. *Nature's Metaphysics*, Oxford: Oxford University Press.

Bognon-Küss, C. 2018 forthcoming. Between biology and chemistry in the Enlightenment: How nutrition shapes vital organization. Buffon, Bonnet, C. F. Wolff. *History and Philosophy of the Life Sciences*.

Cheung, T. 2006. From the organism of a body to the body of an organism: Occurrence and meaning of the word 'organism' from the seventeenth to the nineteenth centuries. *British Journal for History of Science*, *39*(3): 319–339.

Cunningham A. 2003. The pen and the sword: Recovering the disciplinary identity of physiology and anatomy before 1800–II: Old physiology – The pen. *Studies in History and Philosophy of Biological and Biomedical Sciences*, *34*(1): 51–76.

Danchin, E., Charmantier, A., Champagne, F., Mesoudi, A., Pujol, B., and Blanchet, S. 2011. Beyond DNA: Integrating inclusive inheritance into an extended theory of evolution. *Nature Review Genetics*, *12*(7): 475–486.

Fry, I. 2002. *The Emergence of Life on Earth: A Historical and Scientific Overview*, New Brunswick, NJ: Rutgers University Press.

Gilbert, S., and Sarkar, S. 2000. Embracing complexity: Organicism for the twenty-first century. *Developmental Dynamics*, *219*: 1–9.

Godfrey-Smith, P. 2013. Darwinian individuals. In Bouchard, F. and Huneman, P. (eds), *From Groups to Individuals: Evolution and Emerging Individuality* (pp. 17–36), Cambridge, MA: MIT Press.

Jablonka, E., and Lamb, M. 2005. *Evolution in Four Dimensions*, Cambridge, MA: MIT Press.

Ladyman, J., and Ross, D. 2009. *Everything Must Go: Metaphysics Naturalized*, Oxford: Oxford University Press.

Meloni, M. 2016. *Political Biology. Science and Social Values in Human Heredity from Eugenics to Epigenetics*, New York: Palgrave Macmillan.

Nachtomy, O., Shavit, A., and Smith, J. 2002. Leibnizian organisms, nested individuals, and units of selection. *Theory in Bioscience*, *121*: 205–223.

Pigliucci, M. 2007. Do we need an extended evolutionary synthesis? *Evolution*, *61*(12): 2743–2749.

Richards, R. J. 2000. Kant and Blumenbach on the Bildungstrieb: A historical misunderstanding. *Studies in History and Philosophy of Biological and Biomedical Sciences*, *31*: 11–32.

Russell, E. S. 1916. *Form and Function: A Contribution to the History of Animal Morphology*, Chicago, IL: University of Chicago Press.

Walsh, D. M. 2015. *Organisms, Agency, and Evolution*, Cambridge: Cambridge University Press.

Zammito J. H. 2018. *The Gestation of German Biology: Philosophy and Physiology from Stahl to Schelling*, Chicago, IL: University of Chicago Press.

Conclusion

Cécilia Bognon-Küss and Charles T. Wolfe

A well-known German philosopher with his own idiosyncratic interest in biology, Hegel, famously once wrote, *in an introduction* to one of his books, that introductions were not the place to conduct philosophy. It is possible that this moral applies to conclusions as well. This volume, through its different contributions, seeks to investigate the role of a conceptual framework – indeed, several such frameworks – in the coming together of the science we now call 'biology'. It does not claim that 'first there was philosophy, then there was biology'. Indeed, one of the lessons of the chapters written for this book, notably when they treat such figures such as Mme Du Châtelet and Jean Senebier, is that the interpenetration, contamination, mutual influence (choose your metaphor) between a field like 'philosophy' and a field like 'biology' (or physiology, natural history, ecology etc.) is very real and irreducible indeed.

Now, it is true that in the intellectual period *preceding* that which this book investigates, namely the early modern period (the time of Harvey and Locke, of Galileo and Descartes, rather than that of Cuvier, Lamarck or Caspar Friedrich Wolff), something like a philosophy of life (indeed a whole cluster of 'vital matter theories') could be seen to play a role in the investigations into living nature (on this kind of prequel to the present narrative, see Wolfe 2019). But we wanted to focus on the emergence of biology per se: of biology proper. And that sets the historical timeframe squarely in the mid- to late Enlightenment, including what Lynn Nyhart, in her postscript, calls, using a term of Reinhart Koselleck's, the *Sattelzeit* (literally, 'saddle-time', a period in between two canonical periods) of the emergence of biology. So much for the more historical side of the story.

On the philosophical side, one of the points we hope emerges from these chapters is that the face (or landscape, or profile) of philosophy of biology looks rather different depending on which elements of the history of this science it chooses to emphasize. A philosophy of biology in which Darwin is the 'great man' will have different priorities, methods and indeed language than one in which focuses on Kant – or on a network of lesser-known figures who, instead of being 'evolutionists', were fixated on morphology, development and/or laws of vital organization. In his postscript, Philippe Huneman speaks of biology_1 and biology_2 to capture this interaction between 'scientific reflection on living phenomena' and the crystallized science of biology (i.e. biology not as a process or a shifting set of concerns but as a 'thing'). But, interestingly, this distinction, Huneman suggests, also captures the back-and-forth between the historical narrative of biology and the philosophy of biology proper: think, for instance, of the relation between scholarship on the modern synthesis and the philosophical debates of recent decades on the notion of 'fitness'. We'll suggest a related example from a work not so well-known to anglophone philosophers of biology. Georges Canguilhem, in his early study of the development and formation of the notion of reflex action (Canguilhem 1955/1977), tried something similar to what we are suggesting, in perhaps a more blunt way. Namely, he sought to promote the figure of the Oxford natural philosopher Thomas Willis over and against Descartes as a 'father' of neuroscience. Suddenly, vitalistic ideas and chemically determined models of matter were paramount, while brute mechanism – actually, what Canguilhem *saw as* brute mechanism, as one can also be more charitable towards Descartes – receded in the background. Crucially, this was by no means intended as 'just history' (aka 'stamp collecting', in the way some people sometimes put it): Canguilhem was very enthusiastic at being able to quote the eminent twentieth-century neurophysiologist Charles Sherrington's support for Willis, as part of the case he built. He wanted to say that actual neuroscience supported his historiographical revision! In other words, it's not just that all histories of biology (or neuroscience, in this case) reflect the presuppositions and preferences of their authors. Rather, a certain kind of history can have genuine conceptual consequences: in our case, we hope, on certain concepts (or landscapes, or profiles) in and through which biology is reified, operationalized, and projected into the future.

To borrow from a Ukrainian (then American) geneticist almost as famous as Hegel, Theodosius Dobzhanksy, although our line lacks the rhetorical punch of the original: Nothing in biology makes sense except in light of a reflection on how the history of biology and its conceptual accretions interacted with and transformed each other.

References

Canguilhem, G. (1955). *La formation du concept de réflexe aux XVIIe et XVIIIe siècles*, revised edition 1977, Paris: J. Vrin.

Wolfe, C. T. (2019). Life as Concept and as Science. In D. Jalobeanu and C. T. Wolfe (eds.), *Encyclopedia of Early Modern Philosophy and the Sciences*, Dordrecht: Springer.

Index

Printed in the United States
By Bookmasters